SERIOUS PLAY

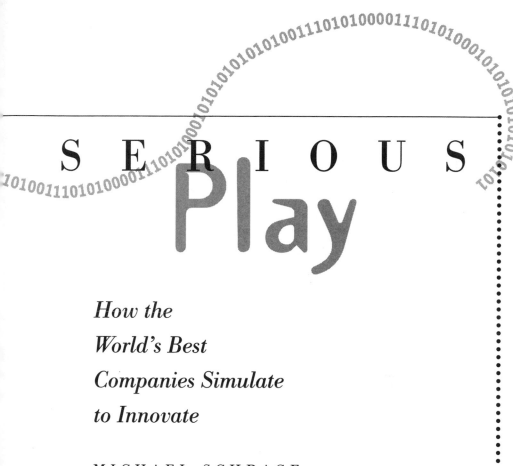

SERIOUS Play

How the
World's Best
Companies Simulate
to Innovate

MICHAEL SCHRAGE

Harvard Business School Press
Boston, Massachusetts

03 02 01 00   5 4 3 2

Library of Congress Cataloging-in-Publication Data

Schrage, Michael.
    Serious play : how the world's best companies simulate to innovate / Michael
Schrage.
        p.   cm.
    Includes index.
    ISBN 0-87584-814-1 (alk. paper)
    1. Technological innovations—Management—Simulation methods.
    I. Title.
    HD45.S368   1999
    658.4'0352—dc21   99-23757
                                                                          CIP

For my mother and her grandchildren

C O  N T E N T S

F O R E W O R D

Research for what became *In Search of Excellence* began twenty years ago. A lot has happened. A lot has changed. And, praise be, a little remains the same.

The bedrock of *In Search of Excellence* were the so-called eight basic principles, the essential practices of the best-run enterprises. Some I'd drop today. Some I'd modify. And one—just one—I'd underscore. Our number one idea has held up! To wit: *"a bias for action."* Mediocre companies, my coauthor Bob Waterman and I observed, pondered and pondered . . . and then pondered some more. (That's what the business schools preached at the time, analysis *über Alles*, "analysis paralysis" per the critics.)

But, we discovered that at the likes of 3M and Hewlett-Packard there was another way. The best label for it was served up to me on a trip to Germany in 1981. An executive from American Express recalled that a former boss's favorite exhortation was, "Ready. Fire! Aim." I loved it. I love it. I put it in our book.

Ready. Fire! Aim. (Dozens have since claimed paternity, including action man Ross Perot.) Truth is Waterman and I were 300 years late to the dance. The scientific method, which helped extract humanity from the Dark Ages, was not built on what has become the B-school analysis paradigm. It was built on . . . *the experiment*, in other words, just do it. Ready. Fire! Aim.

But then along came the "modern" corporation, circa 1850, and its middle managers. And then the business schools followed to codify and certify the corporation's plodding practices. And, ultimately, came *body count*, Bob McNamara's term signaling the apogee (nadir!) of the analytic paradigm, played out in Vietnam.

Bob Waterman and I desperately wanted to change all that. "The Analytic Model Has Led Us Astray" declared one section head in *In Search of Excellence*. Well, twenty years later we have an heir apparent, who has taken our quarter-formed idea and put meat on the bones. I'm being premature perhaps, but I am ready to declare Michael Schrage's *Serious Play* a seminal contribution to the 100-year thread of modern management thought.

In short, I love this book! It is absolutely an original. (And, yes, I'm well aware of how abused the word *original* is.) And it is absolutely right.

Schrage's shtick, *rapid prototyping*, sounds like a third-order innovation tool. Not so, Schrage argues persuasively. Rapid prototyping is the cornerstone, the cultural fountainhead of the innovative enterprise.

The "central thesis of this book," Schrage writes, is that "organizations manage themselves by managing their prototypes." Models, simulations, prototypes all add up to "serious play," which, Schrage claims, "is not an oxymoron [but] the essence of innovation." Serious play, Schragian-style, is no less than the hallmark of 3M, Sony, HP, Microsoft, Walt Disney, and all too few others.

Serious Play is serious work. The power of Schrage's book lies in more than its provocative, profound thesis. The devil, as usual, is in the details—in this case, details that add up to a culture of prototyping.

The prototyping process drives the overall innovation process, Schrage argues in what is possibly the most insightful, counterintuitive twist in a generally insightful book. "Will we get more value from better managing our innovation process to get a prototype?" Schrage asks, "or from better using the prototype to manage our innovation process?" He offers conventional wisdom turned upside down: "Innovative prototypes generate innovative teams." Not vice versa.

The big idea here: the prototyping process becomes the scaffolding for the enterprise's approach to innovation. Through shaping a prototyping process, the firm answers the question, What kind of interactions do we want to create?

There's much more: The physical material used in prototyping is a major strategic innovation variable. For example, the auto industry's long-time reliance on elaborate clay models truncated debate, rather than inviting discussion. Also of paramount importance is who gets to play with prototypes—and when.

I could go on—and would like to. I just want "to be seen" around this set of ideas. Frankly, I'm jealous. I'd love to have written this book! Innovation has long been my passion. My bookshelves groan under the weight of case studies on the creation of television, the radio, the VCR, the Boeing 747, radar, and modern retailing. But *Serious Play* is simply the best book on innovation I've ever read. If you don't read this book and act, if you don't read this book and fundamentally reshape your unit's innovation process, you are making an enormous mistake—and bypassing the number one opportunity to revive and revise your organization.

Read! Act! Now!

Tom Peters
Palo Alto
April 20, 1999

P R E F A C E

I have always enjoyed rehearsals more than performances. Observing a master class in acting or design is almost always more stimulating than seeing the final product. Watching a world-class coach run a practice creates an unparalleled appreciation for the game. Inevitably, I find myself watching and listening for the moments when the right word or gesture changes everything: those moments when talented people play with a possibility, or experiment with a new idea, and *they get it*. The energy and excitement in the group shifts to another level. It's a high. It's fun to watch.

For over a decade now, I've had a relationship with the Massachusetts Institute of Technology's Media Lab, one of the world's premier research centers exploring digital technologies. Its people are wildly smart and passionate. I've witnessed both subtle and more extreme changes in the Lab's culture. One constant, however, is the role of prototypes and play. The Lab's unofficial credo, "Demo or Die!"—the slogan even appears on the face of a fourth-floor clock in lieu of numbers—captures the prevailing belief that it's not enough to have brilliant ideas; you have to be able to demonstrate them. You have to get people to want to play with them. The motto is also sardonic, in that students and professors are constantly being called on to demo for sponsors and visiting luminaries. There's almost always a demo show going on.

The Lab can't be divorced from its demo culture; indeed, the Lab uses its demos to define and refine itself and its future. A successful Media Lab demo isn't show-and-tell; it's show-and-ask. You know you have a really good idea when the reaction to a demo goes beyond "Wow!" or "Cool!" and people start suggesting ways to make the idea better. You know your idea is robust when sponsors can use your demo as a platform for their own work. You know you're doing something special when your demo enlists the brightest graduate students or attracts the most interesting reactions from the Lab's constant stream of visitors.

The best demos inspire their own communities of interest and practice. Some of my best experiences at the Lab have been demo-driven: we would argue and laugh about the possibilities these prototypes presented. We'd bet on which venture capitalists or media giants would fund the next iteration. We'd joke about how to tweak the demo to boost its seductiveness. We'd discuss whether the demo would be more useful as a medium to raise money from sponsors or to inspire new lines of research.

You didn't have to be a sociologist to realize that the Lab's demo culture wasn't just about creating clever ideas; it was about creating clever interactions between people. The best demos let us improvise with each other, not just with the idea. The right word or gesture at the right time could change everything. That kind of creative excitement is addictive. That sensibility was empowering.

These experiences played a seminal role in shaping my first book about collaboration for innovation. I had been fascinated by successful collaborative relationships: Braque and Picasso in the arts, Wilbur and Orville Wright in aviation, Bohr and Heisenberg in physics, Watson and Crick in molecular biology, Bill Gates and Andy Grove in personal computing. I had honestly anticipated insights into collaborative personalities and temperaments, the psychology of collaboration. Instead, I discovered that the key to successful collaborations was the creation and management of "shared space." The notion that more or better communication was the

The Transactional and Collaborative Models
of Communication

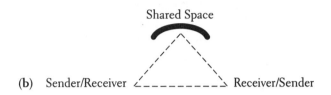

(a) Sender/Receiver ········· Conversation ········· Receiver/Sender

(b) Sender/Receiver   Receiver/Sender

essential ingredient in collaboration was false; what was needed was a fundamentally different kind of communication. The transactional model of communication, which we frequently rely on, is shown in the top half (a) of the figure. The collaborative model is shown in the bottom half (b).

I began exploring the properties of shared space. How could improvements in shared space make collaborations even more effective? My time at the Lab and at MIT's Sloan School of Management had made me particularly sensitive to the role of prototypes as shared space. I was fascinated by the similarities and distinctions between prototyping processes in software development and in new-product development. I was struck by how software designers interacted around software as a modeling medium and how industrial designers behaved around more tangible models. I was particularly intrigued that product design was rapidly becoming more like software design, as industrial designers became ever more reliant on computer-aided tools for building models, simulations, and prototypes. The result? Traditional distinctions between models, prototypes, and simulations were dissolving.

While the tools and technologies of shared space were changing, the (very) human behaviors around them were changing too. Questions of organizational culture in value creation assumed greater prominence. Which made more sense, redesigning the prototype or iterating with key customers? Who got to play with the model, and when, became a mission-critical management issue. Several companies discovered that, instead of innovative teams creating innovative prototypes, as conventional wisdom would suggest, innovative prototypes create innovative teams. I had the good fortune to see these business and design challenges played out at a number of world-class organizations. Invariably, I found that organizations have their own prototyping cultures. Each organization adopted and adapted the new tools for modeling and simulation in its own way. This is not shocking news. What is shocking, though, is that most organizations seem only casually aware that their prototyping cultures often distort the value that their innovation processes are trying to create. I began working with organizations on how to manage their "prototyping portfolios." How could they get the best return on their investments in models, simulations, and prototypes? Should they rethink their cultures of prototyping? How will new-media modeling transform how they innovate with customers and clients?

These are the questions that led to this book. Writing *Serious Play* has turned some of my assumptions and expectations inside out. I came to realize that I had lived my entire professional life under a serious misconception. I had thought of myself as someone who was very good at—and loved to play with—innovative ideas. Give me a clever metaphor, model, simulation, or prototype and I was a cat with a ball of yarn. There was no idea that couldn't be transformed or unraveled simply by batting it around with the right mix of rigor, curiosity, and playfulness.

But as chapters were reviewed and revised, I was forced to confront some simple truths about what I was doing: I really wasn't playing with ideas; I was playing with *representations* of ideas. The notion of talking about "ideas" and "innovations" divorced from the forms that embodied them increasingly struck me as absurd. How I played, how well I played, was overwhelmingly dependent on the nature of those representations. Playing with computer-generated images is as different from playing with spreadsheet software as playing with a football is different from playing with a hockey puck. The nouns shape the verb.

What's more, I wasn't just playing with representations of ideas; I was playing with various versions of representations of ideas. I was constantly shifting between alternative futures and possible pasts. Changing a single word, a single number, or a single command would create new versions for comparison and contrast. Which versions would be managed by my memory? My brain was too small; I needed external versions to really see what was going on. Which versions were explicitly compared and contrasted? Which versions were instantly discarded? How well was I juggling all these various versions?

This process made me painfully aware that I wasn't just playing with these various versions for myself. These models and prototypes were essential to how I worked with others. Innovation was more social than personal. Innovation would be a by-product of how well or poorly I played with others. Behavior—not knowledge, not insight—would drive innovation. How people behaved around different versions of prototypes would overwhelmingly influence how value was created or destroyed.

Consequently, this book has been a journey from ethnography to ethology, from describing cultures of prototyping and simulation to identifying the core behaviors of people who build and use models together. Much of my thinking has been informed by the work of evolutionary anthropologists like Robin Dunbar, Leda Cosmides, and John Tooby, who persuasively argue that our brains are wired as much for manipulation of social relationships as for management of rational thought. Dunbar's thesis that language evolved more as a medium for gossip than for coordinating hunter-gatherer activity is usefully provocative, even if it's only partially accurate.

One can't reflect on how prototypes are designed and used in organizations without wondering to what extent they are products of "arational" human interactions, as opposed to products of rational analyses. That's not accident; that's reality. This book is not a plea for more "rational" management of modeling; it's a call for a greater awareness of the trade-offs organizations must be prepared to make if they want the benefits of prototyping to outweigh their costs. Prototyping is probably the single most pragmatic behavior the innovative firm can practice.

Technology is clearly transforming both the meaning and the media of modeling. Tomorrow's innovators will be evolving prototypes as much as designing them. As Richard Dawkins has so brilliantly articulated in *The Extended Phenotype*, the distinctions between design and evolution in innovation will become harder to assess. What does it mean to be a brilliant designer when synthetic evolution might be a more cost-effective and time-effective way to produce valued innovations? This question is destined to haunt management throughout the next millennium.

The same question will also force organizations to consider what kind of innovators they want to be. At the Media Lab and other world-class organizations, issues of introspection and self-awareness are always bubbling beneath the surface of ingenuity. Serious players are passionate about what they do and why they do it. The works of Donald Schön and Karl Weick are invaluable for exploring how best to integrate organizational and individual introspection. Weick's notion of "retrospective sensemaking" and Schön's "reflective practitioner" are facets of the same phenomenon. One of my most exciting discoveries is that prototyping is guaranteed to make introspection and self-awareness even more important. Kierkegaard once observed that "the sign of a good book is when the book reads you." In the era of digital modeling, the epigrammatic update might be "the sign of a good simulation is when the simulation plays you."

Of course, the most pleasant surprises in writing a book like this are unexpected encounters with creativity and innovation. I never thought my insights into prototyping would be so profoundly influenced by the worlds of theater and animation. Peter Brook, the distinguished English theater director, writes brilliantly of the creative tensions between improvisation and performance in his essay, "The Slyness of Boredom." Anyone who appreciates the thrill of the new can't help but be impressed by Brook's experiences in using improvisation as a tool for innovation. Anyone looking for design inspiration for innovative improvisations will be happy to discover practitioners like Brooks.

My conversations over the years with Disney Imagineers and my reading of Looney Tunes cartoon director Chuck Jones's autobiographies, *Chuck Amok* and *Chuck Reducks,* have offered shockingly useful insight into how bright minds with tight budgets and difficult deadlines have little choice but to play their way to success. Their reliance on models and prototypes — and, even more significantly, on the interactions their models and prototypes stimulate — is of a piece with research at MIT's Media Lab, with product innovations designed at IDEO, and with new financial instruments created at Merrill Lynch. The behavioral similarities are overwhelming. Michelangelo and Microsoft have more in common than not.

Serious play turns out to be not an ideal but a core competence. Boosting the bandwidth for improvisation turns out to be an invitation to innovation. Treating prototypes as conversation pieces turns out to be the best way to discover what you yourself are really trying to accomplish.

In a very real sense, this book has been an exercise in serious play. Ostensibly a finished product, it is also a prototype in process. How so? You haven't yet interacted with it. Please treat this book as an invitation to serious play. Please treat it as a prototype for managing the behavioral challenges of innovation. Like a good prototype or simulation, this book was written to be used, not just read.

A C K N O W L E D G M E N T S

Writing a book often feels less a labor of love than a chronic disease. It's the first thing you think of when you wake up in the morning; it's the last thing that you think of as you're drifting off to a fitful sleep. Writing hurts. The joy comes from finally finishing something that matters. The appreciation comes from looking back and realizing all the people who made the book possible. I am very lucky.

My first thank you belongs to Jennifer Breheny—an excellent researcher who is an even better friend. Her combination of humor, intelligence, and curiosity is responsible for inspiring some of the more unusual insights in this book.

The Design Management Institute's Earl Powell invited me to speak at a 1992 conference, which led to my "The Culture(s) of Prototyping" article in the *Design Management Journal*. That article captured a lot of attention and effectively launched my formal work exploring how organizations build and behave around models. My long-time relationship with MIT's Media Lab provided a wonderful vantage to observe a truly diverse prototyping culture. Nicholas Negroponte, the Lab's director, and Michael Hawley, my old roommate, have been unstintingly supportive as both friends and sounding boards. Much as some musicians have perfect pitch, they have perfect pitch for turning academic concepts into practical prototypes. The support of Tom Malone from MIT's Sloan School was also very helpful.

Ongoing conversations and interactions with Xerox PARC's John Seely Brown, IDEO's David Kelley and Bill Moggridge, Peter Keen, Don Norman, Ikujiro Nonaka, Todd Neal, Cambridge Technology Partners' Thornton May, OneMain.com's Cris Dolan, Rob Fulop, and McKinsey's Brook Manville have all proven invaluable. Special thanks are due to Lazard Freres investment banker David Braunschvig, who has not allowed his talent for facilitating multibillion dollar deals to erode his capacity for creative commentary and critique on how organizations *really* create value.

I am also grateful to Andy Sieg at Merrill Lynch for taking the time and care to connect me to the right people at a truly terrific firm; Anatol Gershman for his support at Andersen Consulting; Shank Hu and Karl Ronn for bringing me into Procter & Gamble; Jim Rutt and Aaron Press for enabling my interactions with Thomson; Alan Naumann and Amanda North at Calico; Jeff Beir and Francois Gossieux at Instinctive; and the folks at CoCreate, NCR, Golden Books, and several other firms that encourage a creative collaboration in prototyping and simulation.

I.D. magazine editor Chee Pearlman ran columns that became the first rough drafts for several passages in this book; the *Red Herring*'s Jason Pontin and Forbes publisher Rich Karlgaard also published my nascent ideas as columns. I was flattered when Alan Webber and Bill Taylor excerpted early portions of this work in *Fast Company*. Of course, I'm thrilled that Rick Tetzeli, Peter Petre, and John Huey at *Fortune* saw fit to bring me on board as their "Brave New Work" columnist as I was finishing *Serious Play*.

The opportunity to present talks on prototyping and simulation at the Product Development Management Association, the Securities Industries Association, Phil Anderson's seminar at Dartmouth's Tuck School, the Corporate Design Foundation, the International Association for Product Design, Procter & Gamble, the International Association for System Engineering, and the Aspen Design

Conference played an essential part in testing and sharpening the focus of my work.

I formally interviewed more than 150 individuals in the course of researching this book. I had casual conversations with at least that many more. Not surprisingly, the chats frequently proved more productive than the in-depth discussions. But my talks with AT&T's David Nagel; Klingenstein, Fields' George Gould; VisiCalc cocreator Dan Bricklin; Lucent's Nobel laureate Arno Penzias; astronaut Jim Lovell; SeeSpace's John Rheinfrank and Shelley Evenson; Sony's Nobuyki Idei; Boeing's Henry Shomber; architect Frank Gehry; venture capitalist John Doerr; Berkeley's Garry Brewer; DaimlerChrysler's John Herlitz and Tom Gale; the Netherland Design Institute's John Thackara; Clark Abt; A.D. Little's Jennifer Kemeny; Disney Imagineers' Bran Ferren, Danny Hillis, and Alan Kay; Gossamer Condor engineer extraordinaire Paul MacCready; Santa Fe's Win Farrell; Yale mathematical economist Martin Shubik; Ziba Design's Sohrab Vossough; Euromoney's David Shirreff; documentarian Karl Sabbagh; Capital Market Risk Adviser's Tanya Beder; evolutionary biologist Richard Dawkins; operations researcher George Gershefski; University of Maryland's Tom Schelling; Stanford's Terry Winograd; USC's Barry Boehm; MIT's Daniel Whitney; John Kao; Microsoft's Mich Mathews; MIT's Lincoln Bloomfield; and MIT architect William Mitchell I fondly recall having had particular impact on my thinking.

My friends—whose wit, charm, and accomplishments are exceeded only by their tolerance—took the time and great care to comment on several chapters. Special thanks to Tom Melcher, Rita Koselka, and Susan Webber, who apparently flashed back to their Harvard B-School days as they ruthlessly critiqued my manuscript. The *Wall Street Journal*'s George Anders, *Business Week*'s David Evans, and Peter Keen were exceedingly generous with their contributions, as was MIT Press's Gita Manaktala and HarperCollins' Lisa Berkowitz. I am grateful that Don Reinertsen and Harvard Business School's Stefan Thomke provided sorely needed reality checks. Pam Alexander, who masquerades as a force of nature when not running AlexanderOgilvy, was a valued friend and confidante during this entire process. Keiko Satoh, John Markoff, and Jeneane Rae also provided welcome support. My brother Elliot, his lovely wife Juliet, and their two boys were a healthy reminder that I needed to play with something besides ideas.

My editor, Marjorie Williams, has been unreservedly encouraging in bringing *Serious Play* to completion. Her ability to give structure to my nonlinear narratives while pushing me to err on the side of boldness was a terrific confidence booster. She has been the biggest booster of this effort from the beginning, and her enthusiasm has been exceeded only by her contribution. I am similarly grateful to Ann Goodsell, whose editorial skills went far beyond traditional copyediting to insightful headlines, commentaries, critiques, and restructurings. Hilary Selby Polk has been a production editor whose abilities to politely but firmly keep me on deadline while making sure that everybody else is adding value to the manuscript at the right places at the right time is quietly impressive. My thanks also to the Press's Genoveva Llosa for her unfailing good humor and effectiveness in keeping the author and his manuscript on the fast track. Sarah McConville's publicity and promotion initiatives are an essential part of why I wanted Harvard Business School Press to be the publisher of this book. The rigorous and fair-minded contributions of the four anonymous reviewers were vital to improving the accessibility of the book you now hold.

When you look at a book at the end of such a writing, review, and production process, you realize with a mixture of incredulity and surprise that it is as much the product of what's been left out as what's been put in. You're haunted by the gaps and seams that only you can see even as you admire the brilliant ways they've been compensated for by world-class editors. You wonder if there weren't—if there aren't—even better ways to express your ideas. Therefore, any and all mistakes and misimpressions are the direct responsibility of the author; the best and most exciting ideas are the direct result of the quality of people with whom the author has had the good fortune to collaborate. These conversations and collaborations are what made *Serious Play* genuinely fun.

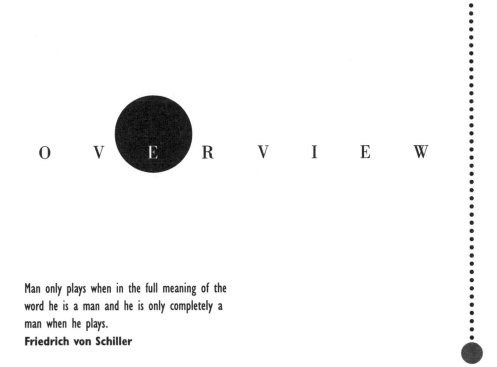

O V E R V I E W

Man only plays when in the full meaning of the
word he is a man and he is only completely a
man when he plays.
Friedrich von Schiller

Serious play is not an oxymoron; it is the essence of innovation. You hear this theme expressed by Nobel Prize–winning scientists like Watson and Crick, pioneering artists like Picasso, and
industry-defining entrepreneurs like James
Watt, Walt Disney, Sam Walton, and Andy
Grove. But what does it mean?

**When talented
musicians improvise,
you don't look inside
their minds; you listen
to what they play.**

It means innovation requires improvisation. It means innovation isn't about rigorously following "the rules of the game,"
but about rigorously challenging and revis-
ing them. It means innovation is less the product of how innovators
think than a by-product of how they behave. Serious play is about
innovative behavior. When talented musicians improvise, you
don't look inside their minds; you listen to what they play. When
talented innovators innovate, you don't listen to the specs they
quote. You look at the models they've created.

The essence of serious play is the challenge and thrill of confronting uncertainties. Whether uncertainties are obstacles or allies depends on how you play. The challenge of converting uncertainty into manageable risks or opportunities explains why serious play is often the most rational behavior for innovators.

Serious play is about improvising with the unanticipated in ways that create new value. Any tools, technologies, techniques, or toys that let people improve how they play seriously with uncertainty is guaranteed to improve the quality of innovation. The ability to align those improvements cost-effectively with the needs of customers, clients, and markets dramatically boosts the odds for competitive success. That is the essential message of this book.

The notion that innovations can be studied or analyzed into existence flies in the face of history and fact. How serious can any study or risk analysis of a proposed innovation be if its authors haven't played with the options, alternatives, and trade-offs? The irony of innovation in any field—but especially in markets wracked by intensifying competition—is that you can't be a serious innovator unless you are willing and able to play. Consequently, serious play shouldn't be treated as either management metaphor or word game. Serious play is both a practice and a perspective on what makes innovation possible. Practically everyone who's been a child knows what it means to play to win, to play for fun, and to play to discover. But this book is not about the business benefits of rediscovering lost childhood; it is about a fundamental transformation in how organizations will go about innovating profitably.

You can't be a serious innovator unless you are willing and able to play.

While the meaning of play remains the same as ever, the toys and games are changing. As the bandwidth for improvisation increases, so do opportunities for richer interactions between people and their ideas. As toys and games for serious play evolve, innovation behavior will evolve right along with them. A new musical instrument may not change the meaning of improvisation, but it will surely change how a talented musician improvises. Similarly, new toys and tools for modeling, prototyping, and simulation may not change the meaning of innovation, but they will surely change how organizations innovate.

Consider time-lapse photography, which enables us to see a flower bloom in seconds, to watch night fall in moments, to witness how a product that takes hours to assemble comes together in an instant. Time compression offers insights impossible to glean from the regular rhythms of natural change. Now invert the point of view: instead of watching the flower blossom, locate the camera within the flower. Focus the camera on all the things happening in the environment that affect how the flower grows: rain, the sun, insects, the gardener. Which perspective reveals more about the flower?

Now, instead of a flower, look at a promising prototype. Position the time-lapse camera inside the prototype to capture its environment. Track all the conversations, collaborations, consultations, arguments, negotiations, debates, budget fights, and brainstorms that go on as the prototype evolves into an innovative product. Which perspective would be more revealing about the culture and behaviors that shape innovation? Which point of view best captures the nature and value of serious play?

This book spends far less time looking at the toys than at the people and organizations that play with them. Behavior change matters more than technological changes. The most intriguing perspectives don't come from looking at what these new toys can do, but from using them to see how people and their organizations behave. Just as, in the words of Leonardo Da Vinci, "The eyes are the window to the soul," toys for serious play can be windows into the soul of innovation and improvisation. There is no better medium for gaining insight into the ethology of innovation than tools for serious play.

Behavior change matters more than technological changes.

Serious Play is divided into three parts: "Getting Real," "Model Behavior," and "S(t)imulating Innovation." Each part contributes a new perspective to the argument that how organizations play with their models determines how successfully they manage themselves and their markets. Each part explores a different facet of what makes models, prototypes, and simulations the most important media for innovation. Jointly, the three parts of the book describe the collision of technology, human behavior, and market forces that will shape tomorrow's management agendas.

Part I, "Getting Real," makes the case that modeling media rewrite the rules for innovation. Chapter 1, "The New Economics of Innovation," argues that changing the economics of play transforms the economics of innovation. More important, when innovation economics change, so do individual and institutional behaviors. Chapter 1 goes on to argue that play is more about behavior than about thought—as is successful innovation. The chapter presents several real-world examples of how organizations successfully use prototypes to redefine their value propositions. Each brief case study looks at how prototypes altered relationships within the firm or with key customers and suppliers. Each offers an actionable insight for managers who want to play with prototyping possibilities.

These examples lead into a more textured case study of how a breakthrough serious play toy—the software spreadsheet—transformed the culture and economics of global finance. Chapter 2, "A Spreadsheet Way of Knowledge," describes the impact of spreadsheets on financial-service firms worldwide and the organizational implications of a modeling tool that massively increased the speed and lowered the costs of simulations and scenarios.

Part II, "Model Behavior," delves deeper into the culture and technology of the marketplace for modeling and simulation. "Our Models, Ourselves" explores the cultures of prototyping and model building in organizations. These artifacts of innovation are products of culture as much as they are products of market forces. Politics, taboos, time pressure, internal strife, and expectations management all play a part in sculpting their value both inside and outside the organization. This chapter outlines the questions that readers need to ask of themselves and their organizations to assess their cultures of innovation and the cultural variables that define how serious play and its toys succeed or fail to effect profitable innovation.

As the digital dynamics of technology shift, so will the challenges that prototyping cultures confront. Chapter 4, "Productive Waste," looks at the complete inversion of historical expectation that innovators face: yesterday they had to manage prototypes as a scarce resource; tomorrow they will have an embarrassment of riches. The challenge of managing abundance is fundamentally different from the challenge of coping with scarcity. Playing with scarcity evokes different improvisations than playing with abundance. "Investing" a portfolio of thousands upon thousands of development cycles demands different skill sets than profitably investing a mere couple of dozen. Where do diminishing returns set in? When cycles are cheap, how many and which new innovation opportunities beg to be explored? Are these opportunities invitations to distraction? Or are they seed capital for tomorrow's innovation?

Playing with scarcity evokes different improvisations than playing with abundance.

It is tempting to assume that a new economy of innovation can lead to more predictable results. But as Chapter 5, "Preparing for Surprise," explains, that's nonsense. Too often, organizations build models and prototypes with the expectation that they will either confirm or crush a pet hypothesis. In reality, models often generate results utterly at odds with the assumptions embedded in them. Models and simulations can be engines of surprise. This chapter looks at why the unexpected results of a simulation may prove far more valuable than the reason the model was built in the first place. Even the simplest simulations can yield counterintuitive insights. The challenge is recognizing and exploiting that unanticipated value. The message is that model surprise may be even more important than model affirmations.

Many surprises aren't pleasant, though, and most cultures of simulation and prototyping aren't healthy. In Chapter 6, "Perils of

Pathological Prototyping," the focus shifts to the dysfunctional aspects of serious play. Serious behavior problems are often associated with serious play and its toys. There is an urgent need to explore where and why model building and playful prototyping degenerate into pathological behavior whose costs consistently outweigh benefits. Managers who see too much of their organizations in this chapter must recognize that the new technologies of serious play represent more of a threat than an opportunity if mismanagement issues are not explicitly addressed.

This point leads directly to the third and final section, "S(t)imulating Innovation." The explosion of new modeling and simulation options—and the risks of prototyping pathologies—raises and renders urgent the question of how organizations can profitably exploit these emerging opportunities. Chapter 7, "S(t)imulating Interventions," addresses this challenge. In many respects, this is a "best-practice" chapter, drawing as it does on the experiences of leading designers, facilitators, and organizations in the realm of innovation. Managing interventions to get greater value from organizations' modeling cultures will inevitably consume an ever-greater portion of executive time and attention. Leveraging—or transforming—prototyping culture will become a perennial top-management priority. Chapter 7 offers actionable insights that increase the chances for cost-effective interventions.

Actionable insights, however, must be linked to the appropriate metrics and measurements. In Chapter 8, "Measuring Prototyping Paybacks," matters of technology, economics, and culture all converge in a series of questions about what kinds of metrics are capable of measuring value even as they facilitate better modeling behavior. Metrics such as "mean-time-to-payback" provide provocative benchmarks against which world-class innovators can assess their prototyping cultures. Indeed, world-class organizations *already* have metrics cultures that both complement and reinforce their vibrant prototyping cultures. The essential point is that culture is not treated as an ethereal, unmeasurable phenomenon: on the contrary, prototypes are sensational management media for stimulating new thinking about and new practices in innovation metrics.

Models, Simulations, Prototypes

Just what are the differences between models, simulations, and prototypes? Once upon a time, there was a fighting chance to answer this question simply and clearly. Today, technologies have conspired to turn any answer into a confusing jumble of semantics that obscure understanding.

Essentially, a model is an approximation of reality that emphasizes some features at the expense of others. A model can be anything from a mathematical equation scribbled on a napkin to a full-scale version of a Boeing 777. The term **model** embraces both simulations and prototypes. It is at the highest level of abstraction.

Typically, a prototype has been a physical model of a product whereas a simulation has been a virtual model of a process. But distinctions blur. Airplane cockpit simulators are physical constructs that simulate flight, and CAD/CAM software allows design engineers to create digital prototypes of the physical objects they want to build.

So is a computer-aided design of a digital automobile a simulation or a prototype? It depends on whom you talk to. At DaimlerChrysler, engineers and management speak of "software prototypes." But when BMW crash-tests its virtual sedans, engineers talk about their simulations. Just as digital technologies have blurred the lines between the physical and the virtual, they have redefined the distinctions between product and process.

In effect, models, simulations, and prototypes have become "flavors" of the same thing—the effort to use technology to re-create some aspect of a reality that matters. Understandably, industrial designers with a love of objects approach the task of prototyping differently from gifted marketing managers concerned about how simulations could create insights into how customers might experience those objects. Different people bring different design sensibilities to this challenge of modeling, simulating, and prototyping.

For the purpose of this book, however, the word **prototype** will carry the load of communicating how organizations use media to manage their innovation processes. While it's not quite fair to say that models, simulations, and prototypes can now be used interchangeably, the distinctions between them are becoming less and less meaningful. The convergence is already having a profound impact on design and innovation management worldwide. That is one of the most important messages of this book.

Chapter 9, "Going Meta: Evolution as a Business Practice," describes tomorrow's innovation environments, in which the collision of technology, behavior, and market forces will be ongoing. Culture will assume an ever-more-prominent role in organizational value creation. Organizations will invest more energy and ingenuity modeling how they model and prototyping how they prototype. The level and intensity of strategic introspection will intensify. At the same time, the accelerating pace of technological innovation will disrupt the traditional paradigm of design and begin to supplement it with the business of evolution. Increasingly, innovations will be managed as products of evolution rather than design. This shift in sensibility promises to further shake up how organizations behave around their innovation infrastructures. While the nature of human behavior won't change any time soon, the challenges that technological innovation poses for corporate culture and human behavior are growing ever more complex. *Serious Play*, both as a book and as a philosophy of innovation, offers a way for individuals and organizations to improvise profitably around the inevitable uncertainties that new media and tomorrow's markets will bring.

> Culture will assume an ever-more-prominent role in organizational value creation.

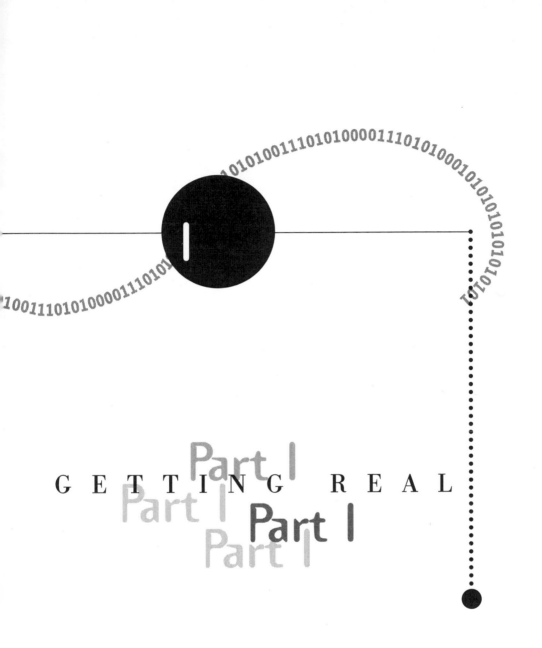

GETTING REAL

Part I
Part I
Part I
Part I

THE NEW ECONOMICS OF INNOVATION

THE ROAD TO WISDOM

The road to wisdom?—Well, it's plain
and simple to express:
    Err
    and err
    and err again
    but less
    and less
    and less.
                Piet Hein*

We shape our models, and then our models shape us.
Virtually every significant marketplace innovation in this century is a direct result of extensive prototyping and simulation. Consider, for example, the airplane, the animated motion picture, the transistor, the microprocessor, the personal computer, the software spreadsheet, recombinant DNA biotechnology, junk bonds, leveraged buyouts, the Internet and its World Wide Web, financial derivatives and synthetic securities, and index funds and yield management.

By shifting from physical clay to virtual clay, every major automobile company has radically reengineered its design and production

---

*Piet Hein © Grook, THE ROAD TO WISDOM. Reprinted with kind permission from Piet Hein a/s, DK-Middelfart.

processes. Boeing's breakthrough 777 jet was built around breakthrough digital prototypes. Walt Disney can't produce feature-
length animations without storyboards. Microsoft could not enjoy
the market share and margins it does without its strategic deploy-
ment of beta-version software. Merrill Lynch's ability to model synthetic securities in simulated over-the-counter markets makes it far
easier to sell its novel financial instruments to Fortune 500 treasuries. Whenever you look for the fundamental dynamics driving
innovation, you find innovators managing models.

New Economics, New Cultures

A new economics of innovation is transforming global business. The computational costs of prototyping products, simulating
services, and modeling business options are shrinking into insignif-
icance. It's becoming ever easier and cheaper and faster to explore
new ideas. "Trend is not destiny," the French poet Paul Valéry once
remarked. This time, I will argue, it is.

Dramatically expand capacity and cut the cost of mechanical
work with Watt's steam engine, and start an Industrial Revolution.
Collect an abundant waste product of the mines—coal tar—to cul-
tivate its myriad properties, and launch a global chemicals and
pharmaceuticals industry. Use circuitry-
etched silicon to annihilate the costs of
computation, and up pop Intel, Microsoft,
Sun, and Cisco.

**A new economics of
innovation is
transforming global
business.**

Quantitative differences create qualitative differences. Make a commonplace
material a dozen times stronger, and it becomes something else.
Make a standard process a thousand times faster, and it becomes
something else. Make a once-scarce resource a million times
cheaper, and it becomes something else. And so do we. When we
radically transform the cost and quality of the raw materials of innovation, we become something else. We start to think twice. We

reperceive. Intuitions recalibrate. Choices shift. We define risk and create value differently. Organizational relationships change. The French industrial sociologist Frederick LePlay brilliantly captured the essence of this transformation in the middle of the nineteenth century: "The most important product of The Mines," LePlay observed, "is The Miner."

In that spirit, consider the possibility that the most important product of innovation is the innovator. That begs critical questions: What kinds of innovators will organizations become as global competition, investor demands, and customer expectations intensify? What new cultures of innovation will emerge as new tools and techniques for prototyping turn the economics of innovation inside out? They will be business cultures that rewrite the rules of enterprise management, entrepreneurship, and value creation. They will be cultures built on the need and desire for *serious play*.

The raw material of innovation

The most important raw material of innovation has always been the interplay between individuals and the expression of their ideas. To paraphrase the science historian Leroi-Gourhan, the evolution of the human mind is basically the evolution of its expressive means. The same thing is true for the evolution of human organizations.

It is still fashionable to assert that managerial minds are possessed by "mental models" that invariably determine what decisions get made. But this is one of those truisms that obscures a larger reality. The mind gets far more credit than it deserves.

As Jay Forrester, MIT's father of systems dynamics, tartly observes: "The mental model is fuzzy. It is incomplete. It is imprecisely stated. Furthermore, within one individual, a mental model changes with time and even during the flow of a single conversation. The human mind assembles **The mind gets far more credit than it deserves.** a few relationships to fit the context of a discussion. As the subject shifts, so does the model. . . . Each participant in a conversation employs a different mental model to interpret the subject. Fundamental assumptions differ but are never brought into the open."

Externalized thought

In order to have actionable meaning, the fuzzy mental models in top-management minds must ultimately be externalized in representations the enterprise can grasp. Visionary speeches and mission statements embodying consultant-certified strategies don't cut it. Mental models become tangible and actionable only in the prototypes that management champions. As Xerox Palo Alto Research Director John Seely Brown puts it, they are "enacted" in the models that managers build.

Prototypes engage the organization's thinking in the explicit. They externalize thought and spark conversation. They're "bandwidth-boosters" and context-creators for both information management and human interaction. A truly effective physical prototype of a portable computer or an automobile dashboard goes beyond the visual to appeal to the tactile and kinesthetic. A genuinely creative spreadsheet simulation of a budget crisis evokes a "suspension of disbelief" that activates the adrenal glands along with the mind. The conversational competitions and design debates that these shared models ignite forge collaborative creativity and fire innovation.

Consequently, these models are not just tools for individual thought. They are inherently social media and mechanisms. More often than not, they become the organization's lingua franca, or medium francum, bridging its multiple Departments of Babel. They boldly engage the "social senses"—the issues of power, perspective, accountability, and control—in ways the firm can't easily avoid.

Collaborative Creativity

Stripped to its essential message, this book is about behavior: how technology behaves, how people behave, and how technology and people behave—and sometimes misbehave—together. The real test of individuals and enterprises is not how much they know or what they know; it's what they do with their knowledge. Making a decision is not the same thing as implementing it.

Today the media and the marketplaces for these expressive interactions are rapidly coevolving in ways that are radically reshaping how people innovate with each other. Tomorrow's innovators will invest more in playing with prototypes, modeling marketplaces, and simulating scenarios because that will become the best way to create new value and profitably deliver it to customers. Innovative models inspire innovative behavior.

Even by accident, new prototyping media can create new interactions between people that in turn create new value. The value of a prototype arises from how productively people interact around its iterations over time. Or, more simply, the value of prototyping arises from how people behave around prototypes.

The articulate prototype

"The hardest single part of building a software system is deciding what to build," observes computer software design pioneer Fred Brooks, author of the classic *The Mythical Man-Month*. "No other part of the conceptual work is so difficult as establishing the detailed technical **Prototypes can be more** requirements. . . . No other part of the **articulate than people.** work so cripples the resulting system if done wrong. . . . Therefore the most important function that software builders do for their clients is the iterative extraction and refinement of the product requirements."

Prototypes, Brooks argues, can be more articulate than people: "It is really impossible for clients, even those working with software engineers, to specify completely, precisely and correctly the exact requirements of a modern software product before having built and tried some versions of the software they are specifying." In other words, creating a dialogue between people and prototypes is more important than creating a dialogue between people alone.

Behavior Matters: The Boeing Example

Boeing's breakthrough 777 jet was the product of a profound transformation of the company's culture and its design technologies. Boeing's traditional hierarchical practice of divvying up the work between disparate engineering departments gave way to new design/build teams. Plans and drawings once sketched on Mylar sheets were digitized and managed by a supersophisticated computer-aided-design program called CATIA.

"We had a saying, 'If it isn't in CATIA, it isn't in the airplane,'" reports Henry Shomber, the 777's chief engineer for digital preassembly. "Integration is what we were really interested in." That desire to integrate represented a radically new mission for the world's largest aircraft company. Boeing's goal, according to Shomber, was to virtually prebuild the entire airplane in CATIA in order to resolve all design conflicts before actual physical assembly. Boeing even devised a digital add-on to CATIA called EPIC (Electronic Preassembly in the Computer) to enable designers and engineers to test how well their virtual components worked together. The company distributed 2,200 terminals to the 777 design team, all connected to what was then the world's largest network of IBM mainframe computers. Key suppliers also had instantaneous access to the data and to changes and modifications as soon as they were confirmed.

Boeing's new digital design infrastructure was so clever that engineers got computer-generated e-mail alerting them to "interferences" created by design conflicts. If the avionics teams and the hydraulics team developed systems that

A large global investment bank realized that its ability to design and deliver innovative synthetic securities swiftly to its big institutional clients depended increasingly on the quality of its software systems. But the bank's financial-engineering "quants"—experts in derivatives design—and its internal software-development group detested each other. The quants believed that the IS department lacked the sophistication and sense of urgency to respond quickly to their design requests. The software developers felt that the quants couldn't prioritize and were too fickle in their systems specifica-

competed for the same physical space in the digital simulation, for instance, CATIA alerted both groups to the conflict. The purpose was to settle conflicts before design prototyping.

Much to their surprise, the 777 project's managers discovered that several engineers deliberately built conflicts with other systems into their proposed designs. Sabotage? Rebellion against the new technology? Engineering humor? Abuse of the prototyping medium? No, the interferences were generated so that engineers in one part of the company could figure out which of their counterparts they should meet with to discuss future design issues. The CATIA-enabled network—created to empower Boeing's engineers and suppliers to build a digital prototype—was also being used to make contact with other parts of the giant organization. In other words, CATIA and EPIC's integrated design and communications system became a tool to seek out collaborations that otherwise wouldn't have existed.

These anticipatory communications allowed people to meet in an unthreatening, exploratory context before difficult engineering disputes emerged. Savvy managers used the network to prevent conflicts rather than manage them after the fact. Several 777 engineers and managers credit this serendipitous use of CATIA for accelerating the pace at which the right people met to solve problems. The prototyping medium generated a new genre of interaction between previously segregated disciplines, transcending Boeing's traditional organizational "silos."

tions. Development initiatives that were to be completed in fewer than 100 days often took six to nine months. Several delivered systems failed to meet expectations even though they largely conformed to the specifications. Each side blamed the other. Management committee meetings often degenerated into shouting matches. These delays and systems failures cost the bank hundreds of millions of dollars in lost profits during a high-growth period in the global derivatives marketplace.

This situation is not atypical. Thousands of companies worldwide waste billions of dollars a year on poorly conceived and poorly implemented information systems. Several global studies estimate that at least one-quarter of internal software-development initiatives are cancelled outright and written off as total losses. Cost overruns, late deliveries, and cancellations are common. Corporate software-development groups often express enormous frustration and bitterness toward their internal clients, whom they characterize as capricious, inarticulate, and clueless about the kind of software they want.

How does this happen? Why should this be? Software development groups, in a sincere effort to be responsive, typically perform an extensive requirements analysis to determine precisely what their clients need, often taking weeks. Users are interviewed at length, and the interviews are translated into systems requirements that are then circulated for approval. Modifications begin. Workflow charts are drawn up and reviewed. Sometimes the serious work doesn't begin until the clients sign off on the requirements analysis.

With client approval in hand, the development group then translates the requirements into specifications for a working prototype. Depending on the project's size, budget, and other commitments, building the prototype can take from thirty to ninety days. This is where the bitterness and recriminations begin. The development team takes their prototype to demo to the client. This is often treated as the single most important meeting in the development process. If the developers are both good and lucky, the demo works well. The prototype not only captures requirements; it anticipates possible improvements. The development team sits back and waits for comments, expecting its rigorous efforts to be recognized and respected.

The client says, "You've given us what we asked for. But it's not what we want."

More often than not, the client responds, "Well, you've given us almost exactly what we discussed. But now that we've seen it, we realize it's not what we really want. We really need you to do something different. How about . . . ?" In essence, the client rejects the prototype. The development team is enraged. The internal client is similarly furious with the developers for being inflexible geeks who don't understand their business. If the project is important enough, the development team sullenly starts all over again, their hostility matched by their client's.

There is a comparatively simple way to avoid this pathology. This approach successfully shocked the troubled global investment bank into a new era of cost-effective collaboration. The key was to change the working relationship by dramatically slashing the time spent on requirements analysis. Rather than gathering requirements from dozens of users, prioritizing, circulating, and seeking approval before initiating prototype development, the team identified the top twenty or thirty requirements as quickly as possible and stopped. Those requirements, and only those requirements, were debated and discussed with the client.

The developers then moved on to rapid development of a prototype. The goal was to present the client with a quick-and-dirty prototype within a fortnight. Why? Because it's far easier for clients to articulate what they want by playing with prototypes than by enumerating requirements. People don't order ingredients from a menu; they order meals. The quick-and-dirty prototype is a medium of codevelopment with the client. Instead of a specification-driven prototype, the result is prototype-driven specs. Clients are equipped to demonstrate and interact with what they want, not merely to describe it. Time after time, this approach achieves productive results by saving time and money while boosting morale.

Furthermore, clients who would be all too quick to toss a development group's hard-won prototype in the garbage are far more reluctant to do so with a prototype that they themselves have helped develop. At the global investment bank, the quants took such pride in the interface design they had helped develop that they packaged it to give to their institutional clients. Quick-and-dirty prototypes can turn clients into partners, enabling developers to manage expectations and deal with changing requirements more responsively.

> **Quick-and-dirty prototypes can turn clients into partners.**

And what gives relentless iteration its great power? "The truth is the clients do not know what they want," Fred Brooks bluntly asserts. "They usually do not know what questions must be answered, and they almost never have thought of the problem in the detail that must be specified. Even the simple answer—'make the new software work like our old manual information-processing system'—is in fact too simple. Clients never want exactly that. . . . So, in planning any software activity, it is necessary to allow for extensive iteration between the client and the designer as part of the system definition."

So much for the crude notion that the customer is always right. On the contrary, without iterative prototyping, it is almost impossible for even the most sophisticated clients to get the software that meets their requirements. Brooks takes the Piet Hein poem at the beginning of the chapter as his design rule: Err and err and err again but less and less and less.

In this sense, the value of prototypes resides less in the models themselves than in the interactions—the conversations, arguments, consultations, collaborations—they invite. Prototypes force individuals and institutions to confront the tyranny of trade-offs. That confrontation, in turn, forces people to play seriously with the difficult choices they must ultimately make.

Prototypes force confrontation with the tyranny of trade-offs.

The fundamental question isn't, What kinds of models, prototypes and simulations should we be building? but, What kind of interactions do we want to create? The latter question aims at the heart of strategic introspection. Consequently, the design focus—the value emphasis—must be on the quantity and quality of human interactions that modeling media can support. Who should be working together? What should they be talking about? Who should see the model next? What is the internal market for value creation? Who "buys" and "sells" prototype enhancements within the enterprise? Who are the "middle men"? Who profits? How will the value created in that market be mapped onto the external marketplace?

Strategic Introspection

The conventional interpretation—in science, academia, and business alike—is that we build "virtual worlds" to better understand the problem to be solved or the opportunity to be exploited. This is accurate without being true. It fails to recognize where the bulk of value may actually be realized. The real reason we need to build and seriously play with prototypes is to get a better understanding of ourselves and our priorities.

The difficult question that organizations must ask is not, What do the prototypes we use tell us about the reality we are trying to manage? but, What do the prototypes we choose tell us about ourselves? The prototypes that organizations generate reflect their own values and perceptions more accurately than they do the realities of the external marketplace.

The Walt Disney Company superbly illustrates how models and prototypes can articulate the internal values of the enterprise. No firm in the world has so assiduously used prototypes to manage its innovation culture and build its global brand equity. "Disney has always been a culture of modeling and simulation," says Disney Imagineering chief Bran Ferren. "The difference is that we use them to create characters and tell stories. . . . We also use them to create experiences." Ferren notes that Imagineers are required to storyboard whatever ideas they propose. Design conversations, he asserts, should take place around designs. Visualizing concepts is even more important than articulating them. As new tools for visualizing design have emerged, Disney has been quick to exploit them. It's no accident that most of Disney's Imagineering Fellows have been leading innovators in computer simulation.

Disney's prototyping culture stems directly from its founder's inner compulsions to model and sketch. According to University of Minnesota art historian and Disney Archives scholar Karel Marling, Walt Disney was obsessed with the third dimension. "Legend has it that Walt Disney never learned to trust blueprints. . . . Like the old-time carpenter who measures rafters by holding every board in place, he believed in things that looked like they really worked,

Mirrorware

Consider a mirror. This mirror is full-length, and you see yourself in it every time you step out of the shower. It can also instantly modify your mirror image in response to voice commands. You can see how you'd look if you lost ten pounds or gained fifteen. If you stopped exercising for two weeks. If you lifted weights three times a week for five weeks. You could see what you'd look like with a deep tan, with a different hairstyle, a different hair or eye color. How you'd look after liposuction. You might enlarge or shrink particular body parts or ask the mirror to age you by three years or ten years. You could ask for best-case and worst-case versions of various aging scenarios. With a Mirrorware upgrade, the mirror can even show you how you'd look in that suit, that tie, that dress, those shoes. What you will look like in that outfit after an hour in the rain. The mirror's projections aren't perfect, of course. So you can store its predictions for comparison with future realities. But the mirror's underlying algorithms can be modified easily to correct for errors or drift.

How much time would you spend in front of the mirror? What questions would you ask it every day? Every week? What would you never ask? What would you ask just for fun? What images would you scrutinize most carefully? How much time would you spend looking at the real you versus images of possible yous? Would the images of possible yous alter how you manage the real you? Would you eat less? Exercise more? Seek surgery? Buy new clothes? In other words, would the mirror significantly change how you behave?

Let's complicate this hypothetical: You have a significant other who cares deeply about you. Would you show your beloved your ideal images? How about the least attractive versions? Would you encourage him or her to request specific modifications of your image? Suppose he or she stored one or two images as benchmarks? Would you collaborate with your significant other to determine the best version of you? Would doing so influence your willingness to pursue it?

models he could examine from all sides, objects he could touch and turn over in his hands."

Marling observes, "When Imagineers go to Disneyland today to

Conversely, would you want to play with the possibilities of your significant other? Or might seeing a best-case or worst-case version create difficult memories for you to manage? How receptive would your significant other be to your hypothetical image modifications? Would such a mirror make your relationship more intimate? Or would mirror imagery become a constant point of reference and contention?

In almost every meaningful respect, this hypothetical mirror is an accurate metaphor for business modeling. Consider a prototype with comparable powers. What questions would managers ask? What questions would be studiously avoided? When a prototype can embody potential as effortlessly as reality, envisioning the future is less a function of knowledge than of desire. As the economist Joseph Schumpeter once observed, innovation is less an act of intellect than an act of will.

Presented with such a mirror, some organizations would muse endlessly on what enhancements would look best; others might look only at their real-time images; still others would alternate between the present image and possible strategic futures. Perhaps a few of the more adventurous would ask their customers and clients to suggest modifications. And a shocking number would never use the mirror at all.

Would such mirrors truly matter? Yes, profoundly. They could utterly transform how organizations see themselves and their futures. They would enable all manner of market differentiation. If we could see our customers and markets in comparable ways, that would be a revolution too. In fact, this is precisely the revolution all organizations are beginning to experience. The fundamental question is not whether or how this mirroring capacity will evolve; it will. The question is, How will organizations get the best possible value from the growing power of these models, prototypes, and simulations?

supervise construction of a Mickey Toontown or an Indiana Jones Adventure, they go equipped with highly finished conceptual illustrations of the end product imagined by its designers, films made with tiny cameras dragged through paper mock-ups of interior spaces, conventional elevations and diagrams indicating the finish desired for each surface." Indeed, Walt Disney's original values continue to live on in current practice.

Prototype portfolios

Every organization has its own array of formal and informal models that it uses to manage itself and its markets. These portfolios of prototypes are, in a sense, investment markets. Managers have to place intelligent bets on which prototypes and modeling processes offer the best chance for creating new customers, new markets, or new profits.

Just as an investment portfolio portrays individual perceptions of risk and reward, prototype portfolios embody the values and priorities of the internal corporate marketplace. Prototypes thus become essential tools and material for managing organizational introspection. This is where their value can have the biggest impact.

More crassly, which prototypes are—to borrow the Boston Consulting Group's famous 2-by-2 matrix—the portfolio's cash cows, stars, question marks, and dogs? Does the organization grasp which generate the best results per resource invested? Does it leverage its modeling investments to better manage itself? Or does the firm avoid hard questions about how well its prototyping portfolio performs? More harshly, *does the firm understand how its internal politics skews the portfolio's returns?*

Prototype portfolios embody the values of the internal marketplace.

These are not mere speculative questions. For Microsoft, Sony, Boeing, GE Capital, Walt Disney, Chrysler, Royal Dutch/Shell, Intel, American Airlines, and other firms that depend on ongoing innovation, managing prototype portfolios is what top management does. There is a reason why Microsoft invests so much in management of its beta-version prototype with key customers and developers. There is a reason why DaimlerChrysler radically reengineered how it prototyped its cars. There is a reason why

Merrill Lynch and Goldman Sachs simulate the behavior of new synthetic securities in collaboration with their largest clients. The reason is straightforward: these prototyping practices yield disproportionately high returns on investment. The hard-dollar benefits can be plausibly measured in the billions. But these practices also disclose a deeper truth. Whenever any of these world-class organizations seeks to enhance its innovation culture and practices, it inevitably discovers that it must significantly change how it prototypes. This can't be overstated. Changing people or tweaking compensation incentives is not enough. Firms have to change how people play with prototypes.

By changing any or all of the variables—the people, the interactions, the versions/iterations, the medium, or the time—an organization can transform, incrementally or dramatically, how it creates value. Change the medium of representation from a two-dimensional sketch to a three-dimensional solid model CAD, and an engineering/manufacturing collaboration is transformed. Insist that industrial designers create five versions of a prototype in six weeks, instead of two iterations in four weeks, and the marketing dialogue expands. Any organization committed to a speed-to-market philosophy inherently commits itself to "rapid prototyping" if it has any hopes for success.

Organizations dissatisfied with the returns on their prototyping portfolios are looking for value through the wrong end of the telescope. They honestly believe that understanding The Problem is more important than understanding themselves. Often, organizations find that their models reveal the gaps between what they claim they *want* to accomplish and what they are actually prepared to *do*.

Modeling the business

As business models themselves evolve and adapt to competition—as product companies like Procter & Gamble and General Motors increasingly wrap services around their goods and as service firms such as Andersen Consulting and AT&T bundle

Modeling the Business: Mexican Ladders

A leading Mexican manufacturer decided to reengineer how it built aluminum ladders. According to its sales and accounting model, the company had been making an operating profit of $4.50 per ladder. The reengineering initiative nearly doubled profits to $8.20 per ladder. Unfortunately, those profit figures proved meaningless. The reengineering had completely ignored the most critical cost issues because the company's business model—and its accounting mechanisms—were flawed.

When the company switched to activity-based accounting to evaluate overhead, its managers were horrified to discover that the company's legal expenses were higher for ladders than for any other product they manufactured. People who fell off ladders tended to sue the manufacturer. Those costs were crippling.

When the total litigation and settlement costs were tallied, the company discovered that it was losing almost $10 on every ladder that it made. And even after the reengineering initiative had slashed manufacturing costs by one third, it was losing more than $6 on every ladder.

Did this company abandon the ladder-building business? No—its new cost model represented an opportunity to redesign its business model. With creative financial modeling and legal finesse, the firm began offering free accident insurance at a cost of just under $2 per ladder. At the same time, the company outsourced all the legal processes associated with compliance, litigation, and claims resolution for a fixed annual fee. Those actions swiftly transformed the manufacturer's large net losses on ladders to satisfactory net profits.

In this case, prototyping ladders and simulating the manufacturing process were precisely the wrong approaches to take. Instead, the company needed to build a model of the **business** of building and selling ladders. By doing so, the company realized that risk-management services offered an innovative opportunity for both cost management and profit. A better model of the business enabled the company to build a business model that profitably married service to product.

In sum, by making it possible to better identify and define the real risks and opportunities of the business, modeling the business can prove far more valuable than modeling its products and services. The premise is simple and compelling: the ways that organizations model themselves, their products, their services, and their markets decisively mold their ability to manage risk and create value.

(Source: Peter Keen, **The Process Edge**)

tangible products into their offerings—new genres of prototypes and simulations emerge. What forms will they take? What new conversations and debates will they provoke? What new value will they enable?

This is as true for Europe and Asia as it is for North America. It is especially true for entrepreneurs who grow entire companies from their prototypes. The cultures that spring up around prototyping and simulation are global business behaviors that transcend borders. They are inherent in every enterprise that innovates in the face of increased complexity and competition. There may be a handful of truly successful global companies—perhaps Philip Morris and Coca-Cola—whose market leadership cannot be directly traced to strong prototyping cultures. By stark contrast, those whose competitiveness utterly depends on their unique cultures of prototyping and simulations include Boeing, Microsoft, General Electric, Gillette, Hewlett-Packard, McDonald's, Sony, IBM, Swatch, Royal Dutch/Shell, Caterpillar, Federal Express, American Airlines, Walt Disney, AT&T, Procter & Gamble, Nokia, Intel, 3M, Merrill Lynch, Toyota, and Detroit's Big Three auto manufacturers. At these companies, the strategies, tactics, and politics of prototyping and simulation frame the strategies, tactics, and politics of value creation.

An organization's culture and prototyping methods effectively determine its ability to profit from innovation. Organizations that can swiftly manage their models, prototypes, and simulations can and do reap tremendous competitive advantage. This has proven true in the past and will prove even more important in the future.

Innovation and Value Creation

The conventional wisdom that "innovation processes" drive prototype development is misleading. Empirical observation of organizations with effective innovation cultures confirms just the opposite: changes in prototypes and simulations drive the innovation process.

Leaders from the CEO on down must step back and ask: Will we get more value from better managing our innovation process to get a prototype, or from better using the prototype to manage our innovation process?

The charismatic prototype

As a rule, managers can learn far more about their company's innovation culture by studying its simulation and prototyping artifacts than by analyzing its espoused innovation processes. Forget traditional management theory: in practice, these modeling media consistently turn the conventional wisdom of innovation management inside out. Management literature credits "product champions" with attracting people to new ideas and new products by the force of their vision and charisma. Tracy Kidder's *The Soul of a New Machine* portrays a mysterious and charismatic leader, West, recruiting talent from all across Data General to create a new minicomputer.

An interesting prototype emits the social and intellectual equivalent of a magnetic field.

Similarly, the influential teamwork paradigm posits that "innovative teams generate innovative prototypes." The corollary is that an organization has to get the right people working together on the right project to create the right prototypes that can be turned into the right products.

On the surface, this reasoning makes sense. Terrific people should create terrific prototypes. But observe how prototypes actually evolve within companies such as Microsoft, Merrill Lynch, DaimlerChrysler, and IKEA, and the conventional wisdom collapses.

Innovation leaders at these companies observe that, more often than not, the precise opposite is the case: innovative prototypes generate innovative teams. The prototype plays a more influential role in creating a team than teams do in creating prototypes. In world-class companies, an interesting prototype emits the social and intellectual equivalent of a magnetic field, attracting smart people with interesting ideas about how to make it better. If you look past the narrative conventions in *The Soul of a New Machine,* a more persuasive—and more accurate—story emerges: the prototype becomes as much of a protagonist as West. This was certainly true of the now legendary invention of the Post-It Note at 3M. Art Fry's desire to benignly mark up his hymnal serendipitously collided with Stuart Silver's prototypical polymer that became the platform for these pervasive yellow stickies. It has also become true at companies like Intel and Hewlett-Packard.

"Once several people have seen an interesting prototype, others hear about it and the researcher gets requests to show off the work," recounts a former Intel research manager. "The use of informal demonstrations facilitates communication between many groups of people. It also assists in getting the world to come to the researchers, rather than placing the burden of communication solely on the researchers." It's increasingly apparent how often people are lured into creative collaborations by "charismatic prototypes"—prototypes that invite participation and enhancement. Charisma, however, has to be placed in both a political and economic context. In fact, market leaders and firms with a particular reputation for innovation can often turn their prototypes into platforms for collaboration. The key is to align opportunities so that customers and suppliers have incentives to co-invest in the prototype's development.

Development is an opportunity for differentiation. Says one former IBM PC designer, "If we don't want our personal computer to become a commodity, then we can't prototype it like a commodity. Unique products come from unique processes; commodity products come from commodity processes."

Relationship management

What are the business implications of this ongoing transformation? Look around. In every industry of global significance, prototyping is assuming greater importance. For world-class companies, there is no longer any meaningful distinction between their prototyping cultures and their innovation cultures. You can't invent

Prototyping Subsidies: Microsoft Example

Prototypes can also be a medium to create and manage a value-added community of customers capable of significantly subsidizing development costs. When Microsoft formally launched its much-awaited and long-delayed Windows 95 operating system in mid-1995, the new software had a huge and immediate impact in the marketplace. Relative to its bulk and awkward design, Win95 was remarkably free of crippling bugs. The operating system had been in the works for years, and Microsoft executives confirmed that Win95 had cost the company "hundreds of millions of dollars" to develop. Microsoft's technical managers understandably took great pride in their rigorous testing and development initiative. Most important, the successful launch consolidated Microsoft's dominant market position in operating systems and greatly enhanced the company's stock market valuation as one of the world's richest technology leaders.

But here's what Microsoft's leaders didn't mention: the world's wealthiest software company had effectively negotiated a billion-dollar subsidy from many of its best customers and developers.

How so? The company had sent out roughly 400,000 beta-version copies of the system to thousands of beta sites worldwide. A "beta version" is a prototype of the final product still amenable to modifications to correct errors and enhance features; "beta sites" are organizations and individuals willing to help track bugs and flaws—and suggest product enhancements—in exchange for receiving the software in advance, influencing its development, and getting help with any problems that might arise. Beta sites are typically lead customers. As a rule, betas tend to be sophisticated corporate connoisseurs of software. The value of their time and technical contributions far exceeds that of naive users.

At first glance, this looks like a good deal for both Microsoft and its beta sites. But even the most conservative back-of-the-envelope calculations reveal an absolutely astonishing transfer of wealth. For the sake of argument, assume that one-quarter of the 400,000 beta versions of Win95 never got tested or were used

the future without first prototyping it. Does anyone believe Disney, Merck, Sony, and their counterparts will invest less ingenuity and care in designing tomorrow's innovations?

in a way totally worthless to Microsoft. That leaves 300,000 beta Win95s on desktops around the world—a sensationally large sample to help Microsoft discover and debug errors, and suggest potentially valuable enhancements. Microsoft drew on a technical population the size of a major city to help improve the quality and capabilities of its new operating system.

A conservative estimate of the cost in time, training, upgrading, monitoring, and maintenance of testing a single Win95 beta for several months is over $3,000. (This estimate is low: estimated total first-year costs of upgrading a corporate personal computer with a finished version of Win95 exceeded $4,000.) Do the math: Microsoft's beta sites invested at least $3,000 times 300,000. Thus Microsoft's global community of betas effectively subsidized the final stage of Win95 development to the tune of, conservatively, $900 million. When one adds the value of customer involvement during earlier prototype phases and of beta sites' suggestions, Microsoft probably reaped well over $1 billion worth of value from its customers and developers before it sold a single copy of Win95. Microsoft's beta sites invested far more in the development of Win95 than did Microsoft itself. Without question, this subsidy enabled Microsoft to produce a far better product for far less money in far less time than it otherwise could have.

Consider how much more competitive an Alcoa, a DuPont, or a GE Plastics might be if they received the equivalent of a $1 billion subsidy from Toyota, General Motors, and Ford to develop lightweight materials for auto bodies. What if Lufthansa, British Air, and Air France provided a $1 billion subsidy for Airbus to build its next-generation airliner? But this subsidy isn't merely about money: it's also about creating and managing relationships that let innovators tap into the time and resources of their savviest customers. The social capital is at least as valuable as the financial capital. Microsoft's proven ability to leverage its prototyping process has given it extraordinary competitive advantages in its global marketplace.

For Microsoft, Intel, and Cisco, the ability to rapid-prototype products with key partners and customers has proven essential to market leadership. Financial-service giants like GE Capital,

Goldman Sachs, and Merrill Lynch have grown dependent on their digital modeling infrastructures for new-product development and risk management. CAD/CAM systems have totally reengineered the innovation infrastructures of global manufacturers like Boeing, DaimlerChrysler, Kyocera, and Sharp Electronics. Meanwhile, ever-faster and ever-cheaper modeling media also create new opportunities for start-up entrepreneurs. The old barriers of entry have been reduced to rubble. As easily as a software spreadsheet can conjure up a business plan, a CAD/CAE program or process simulator can generate a prototype to present to an angel investor or a venture capitalist.

New generations of low-cost prototypes virtually mandate new kinds of innovations in relationships. Thus they guarantee organizations will confront fundamental questions about their internal politics and culture. And, just as important, they give enterprises yet another way of creating proprietary relationships with key customers, clients, and vendors.

Properly managed, prototypes and simulations can actually generate new business models and capture enormous subsidies for the enterprise. They are capable of creating and even managing new kinds of relationships both inside and outside the firm. Models are an even better medium for relationship management than for information management. Prototypes can also transform the speed and nature of communication and collaboration. Multiple representations of a product or service can create a much healthier market for innovation than a single standard.

Models are even better at relationship management than information management.

By altering representations, by managing iterations, by facilitating new kinds of interactions, or simply by changing the time element, disparate organizations have achieved new levels of productivity and effectiveness. By successfully managing their prototyping cultures, these organizations have successfully created new value for themselves and their customers. The lessons here are simple yet powerful. They point to a future in which the ability to manage prototypes creatively becomes synonymous with effectively managing innovation.

The most successful innovators will alternately persuade and intimidate their key customers and suppliers to prototype right along with them. And new modeling media also mean new relationships. Prototypes will become vital links in an organization's "value chain" (and, as the enterprise evolves, essential nodes on its "value webs"). "Modelshare," to coin a phrase, will be recognized as a necessary precursor to mind share and market share.

Choice Creation and Trade-offs

Rehearsal style

Models sculpted from clay differ decisively from models made of bits. They demand different design behaviors, as GM and Toyota have learned. Nokia and GE Plastics would confirm that two-dimensional sketches are not the same as three-dimensional stereolithographic prototypes. And bankers and traders are profoundly aware that spreadsheet software such as Lotus 1-2-3 and Microsoft Excel yields financial models quantitatively and qualitatively superior to its paper predecessors. Changing the modeling medium changes how people manage both the models and themselves.

Indeed, how we think about and play with ideas are grounded in the constraints of our models. As Donald Schön explains in *Educating the Reflective Practitioner,*

> [t]he architect's sketchpad is an example of the variety of virtual
> worlds on which all professions are dependent. . . . Engineers become
> adept at the uses of scale models, wind tunnels and computer simu-
> lations. In an orchestra rehearsal, conductors experiment with
> tempo, phrasing and instrumental balance. A role-play is an impro-
> vised game in which the participants learn to discover the properties

of an interpersonal situation and to reflect-in-action on their intu-
itive responses to it. . . . Virtual worlds are contexts for experiments
within which practitioners can suspend or control some of the every-
day impediments to reflection-in-action.

In addition to conventional prototypes of new products and simulations of new services, an enterprise may model a manufacturing
or customer-support process prior to implementation. It may model
interactions with key customers and suppliers or anticipate hypothetical price wars with competitors. It may even generate provocative scenarios about future market opportunities. What organizations choose *not* to model is as revealing as what they do. What a
coach ignores during practice, what a theater director downplays in
rehearsal conveys priorities as surely as what is stressed. Absence
matters as much as—or more than—presence.

Rehearsal styles, or practice styles, differ. Sometimes a rehearsal
is a vehicle for exploration and improvisation; at other times it's a
rigorous effort to emulate real-world conditions. And some world-
class organizations combine practice and performance seamlessly.
Ballerina Toni Bentley recalls the great choreographer Balanchine's
remark that "every rehearsal should be treated as a performance
and every performance should be treated as a rehearsal." That is a
compelling design philosophy.

If the hard-dollar and time costs associated with rehearsals were
slashed, would organizations invest more time and effort in exploring options they had never before considered? Or would they spend
more effort testing the implications of their favorite options? Are
there sound reasons why an organization might never model particular tasks, no matter how cheaply or easily it might be done? These
questions evoke unease. Do some choices reflect cultural pathologies? Or do they create comparative advantage in an organization's
internal economics of modeling? Honest answers can give organizations the insights necessary to dramatically improve their innovation infrastructures.

Tensions occur as the behavior of models coevolves with the
behavior of people. These tensions are certain to increase. Indeed,

their intensity may prove almost unbearable. Why? Because, as prototyping media grow ever more powerful and ever cheaper, it becomes far more difficult for managers to resist the challenge of exploring just one more design option . . . of reducing a risk exposure a few more percentage points . . . of testing that one shaky assumption . . . of collaborating with a key customer in just one more iteration. Where does that line get drawn? Who gets to draw it?

Prototyping trade-offs

In the real world, prototypes quickly become tools for negotiating trade-offs. Should the process be optimized around minimizing costs or maximizing output? Which features must be sacrificed to reduce costs 20 percent? What is the warranty exposure for adding this capability? Does this functional upgrade justify an estimated 7 percent degradation in overall performance? Form or function? Design for manufacture or design for maintenance? Should we ship with 85 percent finished to hit the market ninety days early? Or wait ten weeks to get quality up to 98 percent?

Essentially, the purpose of a prototype is to create or unearth choices. The best prototypes are those that produce the most useful and important choices. They in turn define the context for trade-offs. Prototypes can be a surrogate marketplace around which economic value gets created and traded. Indeed, modeling media are typically more cost-effective than marketplaces for pinpointing trade-offs that ought to be explored. Sometimes, the answers are self-evident. Not infrequently, those self-evident answers are ignored. But make no mistake: these trade-offs—and the negotiations and budget battles that surround them—determine how organizations manage themselves and their innovations. They organize how firms play with possible realities. They are how successful organizations get real. As technology transforms the economics of modeling media, it will increasingly force firms to

> Prototypes are machine tools for producing choice.

reconsider their economic choices and behaviors. As prototypes become ever more powerful and persuasive, they will compel new intensities of introspection. To paraphrase philosopher Alfred North Whitehead, they will become conceptual machine tools for postindustrial innovation—not because we are now gifted with finer imaginations but because we have better instruments for imagining and rehearsing the future. These opportunities to model the future world swiftly, cheaply, and creatively will equip organizations to bridge the gaps between innovative ideas and innovative behavior.

The dominant theme remains behavior: How do people behave around prototypes? How will business behavior change as business models change? Fifty years of technological revolution certainly don't trump 3 million years of biological evolution, but technologies do increasingly mediate human experience. How can that fact create new value for the enterprise and its customers? Clearly, the most important questions about the future of any firm will revolve around human behavior. That's why managers shouldn't just ask, What will this model do? but, What will this model do to people?

I will argue that, outside of the marketplace itself, there is no better way to gain insight into individual and organizational behavior and to promote value-creating behavior than through the use of prototypes. They will become an ever more important investment in the quality of an organization's human capital and its innovation capacity. The smartest, most creative people in innovative organizations already spend more time interacting around more versions of models of their most important products and services than ever before. Tomorrow, they will spend even more time. But will that time be spent more productively? How will they—and their investors—know?

A SPREADSHEET WAY OF KNOWLEDGE

Who remembers why Interco died? Interco, which owned some of America's best-known consumer brands—Converse sneakers, Florsheim shoes, Ethan Allen furniture—crumpled into bankruptcy beneath $2 billion of debt in the wake of a recapitalization initiative. The company had become the willing captive of software spreadsheet calculations that stretched credulity beyond the breaking point. At a board meeting in 1988, the investment bankers presented twenty-six pages of complicated spreadsheets demonstrating that, at worst, Interco's shareholders would receive a 20-percent return on the recapitalization plan. The numbers were never credibly challenged. The board believed the spreadsheets. The company's fate was sealed by an overdose of toxic software.

The story of the financial innovations and financial engineering that transformed the business landscape of the 1980s is written in the cells of spreadsheet software. Every major deal was touched by this technology. From Michael Milken's high-yield "junk-bond" financings to Kohlberg Kravis Roberts's leveraged buyouts to the global explosion of "synthetic securities," spreadsheets functioned as the computational catalysts accelerating and intensifying the dynamics of deal making. Venture capitalists, investment bankers, and mutual-fund managers enthusiastically embraced them for their ease and power. Throughout the spectacular changes in global finance during the 1980s and 1990s, power and influence flowed from the matrices of software spreadsheets.

"Spreadsheets totally changed the financial business," observes George Gould, a cofounder of the Donaldson, Lufkin, Jenrette investment-banking firm and undersecretary of the Treasury in the Reagan administration. "Certainly, spreadsheets made CFOs more powerful than they used to be—a fact that is reflected in their pay scales."

Models that once cost thousands of dollars now cost pennies.

Low-cost spreadsheet software effectively launched the largest and most significant experiment in rapid prototyping and simulation in the history of business. The culture of finance had never seen a tool like it. On both a relative and an absolute basis, spreadsheet software proved astonishingly fast, easy, and cheap. Financial models that had once cost thousands of dollars to design and build now cost thousandths of pennies. The marginal cost of modifying complex spreadsheet models plummeted to near zero. Moreover, the technology was culturally compatible with the mindset of finance. Using spreadsheet software became easier than not using spreadsheet software. Within five years of the 1979 introduction of VisiCalc, the first electronic spreadsheet for personal computers, over 1 million software spreadsheets were being sold annually. Within a decade, North America's financial culture had become a spreadsheet culture. Spreadsheets became the medium, the method, the tool, and the language of all serious financial analysis.

The Impact of Spreadsheets on Financial Organizations

More important, spreadsheet culture profoundly altered managerial behavior in the financial world. The artfully crafted spreadsheet—Here's where we can cut costs! Here's how we should restructure this deal!—could and did prove as politically incendiary a document as the Declaration of Independence or the Communist Manifesto. Spreadsheet software breathed new life into the old adage, "Figures don't lie but liars figure." As budgets and forecasts were used to find previously unimagined opportunities, traditional perceptions of power, politics, productivity, and profit all dramatically shifted. Finance-driven organizations often found themselves reorganizing around their spreadsheets. This phenomenon was particularly widespread at investment and commercial banks and the corporate treasuries of Fortune 100 enterprises.

Operationally, Gould asserts, spreadsheets affected every significant facet of finance. "They were the great leveraged-buyout tool of that [1980s] era," he notes. "They turned what had been a traditional financial analysis into a blueprint of how to run the business to maximize cash flow. Mergers and acquisitions once driven by long-term investment-banking relationships were now being driven by aggressive young bankers with even more aggressive spreadsheet models. But they were seen as credible models, so boards of directors were legally obligated to take them seriously."

The ascendancy of give-and-take

Gould—a traditional investment banker, who recalls the pre-software-spreadsheet era with more than a trace of nostalgia—observes that spreadsheets totally transformed the nature of discourse in the field. "We went, really, from almost no ability to present financial alternatives to a very dynamic form of interaction where all we do is talk about alternatives," he comments. "We've gone from very little give-and-take around paper ledgers to the expectation that a conversation around a computer spreadsheet model will be almost entirely give-and-take."

A conversation around a spreadsheet is almost entirely give-and-take.

Indeed, Gould recalls some of the more contentious conversations surrounding the government's multibillion-dollar supervised sale of distressed savings-and-loans to private investors as battles between dueling spreadsheets—battles, he ruefully concedes, in which government negotiators may have been "outspreadsheeted" by their private-sector counterparts. "They would often bring their personal computers to the negotiations," says Gould, "and they could model practically every disputed issue instantly."

The accelerated tempo of finance further changed the rules of the game. Financial innovation resembled less a game of grand-master's chess than a dollar-denominated form of Space Invaders in which skillful and speedy improvisation acquired parity with penetrating analysis. The ability to generate scores of useful spread-sheet scenarios overnight became as valuable as creating one "right" spreadsheet model. Forecasts and financial innovations could be prototyped with astonishing speed and comprehensive-ness. Michael Milken, for example, has privately credited spread-sheet software with making it possible to model and market so swiftly "high-yield"—aka junk—bonds he pioneered at Drexel Burnham Lambert.

Spreadsheet cultures

Of course, different organizations employed spreadsheets differently. Inevitably, spreadsheet software became as much an artifact of corporate culture and the internal economics of the firm as a tool for rational financial management. Both by default and by design, financial-service firms rapidly evolved their own spreadsheet-based cultures. The larger investment banks—Merrill Lynch, Morgan Stanley, Goldman Sachs, and the like—typically tossed platoons of young analysts at deals to churn out reams of spreadsheets. Their premise was that quantity would eventually beget the appropriate quality; large-scale "number crunching" was the necessary grunt work to get the deal done. At first, managing directors at these firms seldom saw spreadsheets on-screen, prefer-ring instead to look at the printouts their MBA underlings brought them. At the toniest *banques d'affaires* and trading houses, by con-trast, where computational sophistication and top management were not mutually exclusive, spreadsheets became strategic instruments for customizing deal structures and crafting swaps. These firms treated spreadsheets more like mission-critical machine tools than mere media for staff support.

An explosion of questions

The rapid diffusion of spreadsheets forced financial-service companies of all shapes and sizes to confront ethical and operational issues with new urgency. Should firms insist on internally consistent assumptions in the spreadsheets they presented to clients? Or should kinks in key financial assumptions be subtly smoothed by the software to get the deal done? Should certain spreadsheet models be proprietary to the firm? Or should they be given to wary clients to help persuade them of the accuracy of key analyses? Did it make sense to use spreadsheets primarily to automate existing financial analyses? Or might there be more bang for the buck using them to sniff out innovation opportunities? What were the trade-offs? How might they best be calculated?

Spreadsheet software enabled organizations to ask themselves questions they had never been able to ask before. And the same spreadsheets could be used to help answer those questions.

This ongoing global experiment affirms the hypothesis that the proliferation of cheap prototyping tools can dramatically transform how an industry manages itself and the value it creates for others. A corollary hypothesis is that an organization's ability to create value now depends on its ability to use these tools effectively. The business issue here is emphatically not the conventional challenge of better information or more effective knowledge management. It is, instead, what kind of prototypes must the organization create in order to create new value? How does an organization's culture of prototyping correlate—both positively and negatively—with its ability to create value?

> **Spreadsheet software raised questions that firms had never been able to ask themselves.**

Lessons of Spreadsheet Proliferation

The impact of spreadsheet software on financial services thus offers a powerful case study on the challenges posed by radical innovation in prototyping. Global finance represents a near-perfect microcosm of the array of business opportunities—and cultural pathologies—that organizations invariably confront when new modeling media proliferate. As Benjamin Franklin once observed, "Experience is the best teacher but only a fool always insists on learning from her." The spreadsheet experience teaches vital lessons about managing the hard task of serious play.

Business opportunities proliferate along with modeling media.

What are those lessons? What do they teach? Spreadsheet history offers an arsenal of themes that potentially apply to all manner of prototyping media and organizations. Later we will look at the "critical success factors" for such media. The lessons of experience that follow offer a context for assessing how to get the maximum value for time, money, and resources invested in making models that matter.

Immediate payback

Dan Bricklin, the Harvard Business School student who created VisiCalc with MIT's Bob Frankston, attributes the success of his software to the speed with which it paid for itself. Bricklin observes that well-heeled Wall Street analysts—thoroughly sick and tired of recalculating spreadsheet after spreadsheet on paper—would cheerfully shell out over $2,500 to buy VisiCalc and an Apple II personal computer simply to be able to reduce the time and tedium associated with the manual approach. "For most of these guys," Bricklin recalls, "the payback for their investment was under a week." Casual conversations with early adopters confirm that they believe their purchase paid for itself within a fortnight— including time spent learning to use the software proficiently. Market-research data also shows that, throughout the early 1980s, most finance-oriented professionals purchased personal computers expressly for the purpose of running spreadsheets. Don Estridge, the late IBM executive who oversaw the computer giant's breakthrough personal-computer business, freely acknowledged that spreadsheet software, not word processing, was the number one reason why people purchased IBM's earliest machines.

> **Spreadsheets paid for themselves faster than any other productivity tool.**

Mitch Kapor, the entrepreneur who launched Lotus 1-2-3 — the spreadsheet-software package that became an industry standard — agrees that swift "mean time to payback" was a crucial reason for its success. For the first five years of the spreadsheet's existence, he notes, individuals — not institutions — were responsible for the bulk of sales. Ultimately, larger enterprises came to accept what individual experience had confirmed from the beginning: spreadsheets could pay for themselves faster than practically any white-collar productivity tool of the time. They were perceived as a terrific investment.

Seamless automation and speculation

It is noteworthy, however, that VisiCalc, Lotus 1-2-3, and Microsoft Excel were initially purchased less for their power to prototype financial futures than for their ability to automate the arduous task of spreadsheet computation. It's almost impossible to overstress the appeal of a modeling medium that offers not just one but two compelling value propositions. Dan Bricklin's motivation to invent VisiCalc was precisely to avoid the pain of pencil-and-paper ledger processing. As technology writer Steven Levy recounts in his seminal 1984 *Harper's* article, "A Spreadsheet Way of Knowledge," Bricklin's Harvard Business School class had been asked to

> *project the complicated financial implications — the shift in numbers and dollars, and the shifts resulting from these shifts — of one company's acquisition of another. Bricklin and his classmates would need ledger sheets, often called spreadsheets. . . . The problem with*

*ledger sheets was that if one monthly expense went up or down every-
thing—everything—had to be recalculated. It was a tedious task,
and few people who earned their MBAs at Harvard expected to work
with spreadsheets very much. . . . Bricklin knew all this, but he also
knew that spreadsheets were needed for the exercise; he wanted an
easier way to do them.*

VisiCalc (for Visible Calculation) was Bricklin's breakthrough
solution to this problem. But VisiCalc's brilliance transcended
automation. The same technology that slashed hours from budget
recalculations also enabled its users to build alternate spreadsheet
realities cheaply. "What really has the spreadsheet users charmed,
wrote Levy, "is not the hard and fast figures but the 'what-if' factor:
the ability to create scenarios, explore hypothetical developments,
try out different options. The spreadsheet, as one executive put it,
allows the users to create and then experiment with 'a phantom
business within the computer.'"

In other words, VisiCalc and its successors created a seamless
continuum between automation and speculation. There was no difference, technologically speaking, between modeling and forecasting imagined numbers and manipulating real numbers. The
spreadsheet didn't care. The same learning that went into planning
a real budget could be used to craft a new business plan. The
spreadsheet became a medium for serious play as much as for
crunching hard numbers.

Simultaneous risk management and opportunity creation

The same spreadsheets that let analysts track expenses and
hedge exposures also empowered innovators to test ideas and craft
new products. Wall Street "quants" used spreadsheets as virtual
"wind tunnels" to test-fly the robustness of their proposed innovations. "Of course, we still used mainframes back then," recalls a former Salomon Brothers derivatives partner who enjoyed several
years of million-dollar bonuses for his work. "But we could rapid-
prototype a dozen flavors of a new product on our spreadsheets in

an afternoon. We could redesign stuff on the fly before we took it to the big machines for heavy-duty testing."

But a critical convergence happened along the way. Creative risk managers at many financial-service firms realized that spreadsheets created new windows to exploit costs and risks. Expenditures could be transformed into investments, purchases could be restructured as leases, and, with a little ingenuity, cost centers could be transformed into profit centers. Aggressive insurance companies like AIG and reinsurance companies like Swiss Re used spreadsheet analyses opportunistically to devise new financial products for their traditional risk-management sectors.

> **Wall Street used spreadsheets as wind tunnels to test-fly innovations.**

Conversely, financial innovators discovered that options and futures could be redesigned to offer a new generation of hedging techniques. Investment innovations like "portfolio insurance" are attributable to opportunity creators who appreciated the market potential of risk management. In other words, spreadsheets allowed for an unprecedented integration of risk management with opportunity creation. Risk managers increasingly used spreadsheets to become more opportunistic; financial innovators increasingly used spreadsheets to better manage the risks associated with their offerings.

Integration/fusion of analysis and operations

America's top leveraged-buyout firms quickly turned spreadsheet software from an analytical tool that could model the business into computerized control panels that could help run it. Spreadsheet-driven PCs became like supercharged x-ray machines capable of not only seeing broken bones but also helping set them. Spreadsheets equipped managers to fuse financial diagnoses with possible prescriptions and treatments. Paper spreadsheets could statically track cash flows; software spreadsheets could dynamically manage them. "They are how we ran our businesses," says Kevin Bousquette, a former Kohlberg Kravis Roberts associate.

For traditional investment bankers, spreadsheets dissolved the conventional distinction between analysis and operations in their multibillion-dollar mergers-and-acquisitions (M&A) business. During the go-go merger era of the 1960s, for example, white-shoe investment bankers typically relied on a blend of long-term relationships and exhaustive financial analysis to position a merger offer. During the 1980s and 1990s, the flexibility and dynamism of spreadsheets turned these analyses into marketing-and-sales vehicles that could custom-fit deals.

The frenetic M&A auctions environment of the late 1980s created contests in which, as soon as the terms of a deal were revealed, rival bidders would instantly crunch the numbers to pinpoint weaknesses and counter with better terms. A former First Boston associate from the aggressive Wasserstein and Perella M&A team recalls churning out dozens of spreadsheets overnight as ardent suitors sweetened their acquisition offers. Instant analyses turned into prototypes for instant counteroffers.

Spreadsheets became analytical and operational media for designing, negotiating, and closing deals. The subtext of the takeover of RJR/Nabisco—by far the biggest leveraged buyout of the 1980s, recounted in *Barbarians at the Gate*—is that the ultimate purchase depended less on the charm or deep pockets of any particular bidder than on the combined analytical and operational persuasiveness of their spreadsheets. A charismatic spreadsheet could prove quite as charming as the most charismatic investment banker. Indeed, there was nothing quite like a persuasive spreadsheet forecast to make a charismatic banker even more compelling.

Models as a medium of communication and collaboration

In finance, spreadsheets have undeniably come to define the deal, the plan, and the budget. The multinational, multibillion-dollar industrial leader Asea Brown Boveri, nominally headquartered in Europe, insists that English is its primary language, but in practice the dominant language of ABB is spreadsheet. Indeed, several managers observe that it is communication, disagreement, and negotiation over spreadsheet forecasts and projections that drive dialogue at ABB. "Yes, I think it is better . . . to be more fluent in your spreadsheet forecasts than in English," says a senior ABB executive. "Our numbers are probably more important to us than our words."

This sentiment, echoed at many financial-service firms, underscores a central point about the evolution of spreadsheet modeling: What began as a personal tool has become a social process. Spreadsheets are as much about communicating ideas and persuading colleagues and clients as they are performing credible calculations about plans and budgets. "People doing negotiations now sit down with spreadsheets," says VisiCalc cocreator Bob Frankston. "When you're trying to sell a car, the standard technique is to ask for the other person's objections and then argue them away. If two people are in front of a spreadsheet and one says, 'Well, the numbers say this,' the other can't say, 'Yes, but there's something I can't quite point to.'"

Top-tier financial firms like Merrill Lynch and Goldman Sachs often sit down with their clients to collaborate on the design of new financial instruments and trading strategies. The spreadsheet model is the "shared space" where ideas are created and their practical value is debated. The simulation ends up managing expectations as well as communications. The spreadsheet projections present a portrait of a predicted future. Spreadsheets can speak louder than words.

A spreadsheet model is "shared space."

Just as rhetoric marshals words into persuasive arguments, there is also a rhetoric of spreadsheets that permits numbers and extrapolations to be blended into persuasive forecasts. Spreadsheet rhetoric is about turning what appears to be the dispassionate logic of numbers into presentations that mute opposition even as they enlist allies. The cells and the equations that drive them become the nouns, verbs, and adjectives that shape the very scenarios they purportedly describe. Spreadsheets aren't just financial models; they are financial metaphors. Metaphors are rhetorical devices, designed to persuade. Increasingly, the rhetoric of spreadsheets has become the means by which organizations are persuaded to innovate and cut costs.

Ease of linkage to other prototypes

Back in 1984, strategic planner Peter Schwartz of Royal Dutch/Shell felt frustrated by the inability or unwillingness of the giant oil company's management committee to address troubling scenarios the company's planners had devised. So Schwartz reframed his scenarios by connecting them to software spreadsheets. Schwartz whipped up a simple computer model and linked it to a Kodak Ektachrome projector in a bid to get the managing directors to consider the possibility of an oil-price crash. When the meeting began, the managers resisted. "We don't play with models in the boardroom," one manager declared. So Schwartz began playing with the model himself. Gradually, the managing directors began to speak up. After an hour, they were so intrigued that they scheduled a follow-up meeting. "They couldn't leave," Schwartz recalls. "They were totally hooked." This jury-rigged presentation convinced Shell's top leadership to rethink how to manage the company's future options.

Though Royal Dutch/Shell's strategic planners took justifiable pride in their narrative-style scenarios, the business reality was that spreadsheet projections added a necessary quantitative spine to their conceptual bones. The scenario illuminated—and challenged—assumptions embedded in the spreadsheet; conversely, the spreadsheet pushed the boundaries of the scenario's narrative. The basic point, however, is that the formal scenario became more valuable linked to the spreadsheet, and the spreadsheet became more important linked to the scenario.

Whether as interface or complement, integrating spreadsheets into other models virtually guarantees them an even more central position in the firm's strategic conversations and creative collaborations. Spreadsheets are emphatically not stand-alone simulations. As modeling media proliferate, we will see a growing coevolution and hybridization of modeling and prototyping media.

The challenge of "surplus" spreadsheets

VisiCalc was a breakthrough innovation that genuinely lived up to economist Joseph Schumpeter's definition of innovation as "creative destruction." Software-spreadsheet computation quickly brought paper-ledger processing to the brink of economic extinction. Financial models that had cost thousands of dollars and taken days to produce now cost thousandths of pennies and take milliseconds. And cheaper spreadsheets invariably meant more spreadsheets: A spreadsheet population explosion resulted. Thus, a virtue gave rise to an unseemly vice: organizations began to drown in their own spreadsheets. In some companies, the new abundance became an excellent excuse for "analysis paralysis." There was almost always a good reason to produce "just one more spreadsheet."

A boon gave rise to a vice: organizations began to drown in spreadsheets.

The new spreadsheet economy changed cultural expectations in the world of finance: If it could be modeled quickly and cheaply, why not do it? Quantity of spreadsheet analysis became as important as quality. For every finance-oriented question, the organization increasingly expected a spreadsheet answer. Lotus Development's Mitch Kapor lamented that he had created a software monster: people in his own company were spending more time on mindless calculation than on genuine thought. "We had managers producing spreadsheets because they could," Kapor recalls. The ability to generate spreadsheets on demand completely changed how organizations reviewed their forecasts. At one consumer-products giant, brand managers seized on spreadsheets as a weapon to fend off harsh senior-management interrogation of their budgets. As personal computers swept through the company in the mid-1980s, brand managers who had spent dozens of painful hours producing budget presentations giddily embraced the computational riches their spreadsheet software conferred. Budget-review committees accustomed to presentations consisting of a handful of spreadsheets now confronted budget documents thicker than the federal budget. Ambitious and eager-to-impress young brand managers calculated budgets anticipating every possible market contingency. Top managers increasingly found themselves examining spreadsheets rather than plans.

Within eighteen months, senior management had had enough, and the rules of the budgeting game changed: brand managers were told that they could present only three spreadsheet models for review—Best Case, Worst Case, and Most Likely Case. In other words, senior management imposed constraints—regulation—to force the brand managers to better manage the spreadsheet explosion. The result? Brand managers had to develop thorough rationales for each of their spreadsheet scenarios, rather than contingency scenarios for every possibility. Top managers in turn could better weigh the assumptions and computations underlying the three cases.

The point is that the new economics of spreadsheet software had fundamentally shifted the "point of diminishing returns" for time invested in spreadsheet modeling. Just how much more valuable was the thirtieth spreadsheet than the twenty-fifth? At what moment did new spreadsheet iterations begin to yield diminishing returns? Who or what determined that point? A boss? Peer review? A deadline? Did organizational culture, politics, or economics define diminishing returns? Similarly, how many types of contingencies did managers need to plan for? In the paper-oriented past, the limitations of the medium had restricted the number of spreadsheet forecasts that could be computed by deadline. In the post-VisiCalc era, technical limitations vanished; the remaining constraints were time, managerial interest, and the perceived value of the models.

Abundance made organizations more appreciative of scarcity.

The ability to create so many quality spreadsheets so quickly and cheaply generated an ironic productivity paradox: all that abundance made organizations more appreciative of scarcity. What number and types of spreadsheet models would generate the most value for the organization? Before 1980, this question would have been bizarre. By 1990, it had become a mission-critical challenge to every single financial-services innovator competing in the global marketplace. As the cost of spreadsheet iterations has become marginal, organizations have had to explore "regulatory" and "market" mechanisms to manage their spreadsheet economies.

Cheap computation, expensive assumptions

"Because the marginal cost of computation drops to close to nothing, people who are not particularly good at modeling models can model like crazy," says Martin Shubik, an applied mathematician at Yale and game-theory pioneer who has consulted on simulation to the Pentagon and industry. Noting the now-commonplace phenomenon of dueling spreadsheets, Shubik observes, "There are big software spitball fights because my expert brings bigger and faster simulations to the table than your expert."

> Spreadsheet projections are true to their underlying assumptions.

The catch is that spreadsheet projections—assuming computational integrity—will be true to their underlying assumptions. This means that debates about organizational strategy and tactics shift from the technical validity of the projections to the merits of their assumptions. In other words, as the ability to model becomes more of a commodity, the most compelling issues become the assumptions on which those models are based.

"The prevalence of the computer and the spreadsheet has caused entrepreneurs to develop pro forma projections of their businesses," observes Don Valentine, the venture capitalist and early investor in Apple Computer and Cisco Systems.

Sometimes you find incredible market shares achieved in just four years, simply because the numbers suggest it; or sometimes you find sales expenses projected at numbers far below what you'll probably need to get the orders. Things like that fall out because the numbers

*are generated impersonally, and nobody thinks about the relation-
ship between them and the structures that have to be in place before
the numbers start coming in. . . . So for us the business plan is a
bridge to communication. . . . We've become less concerned about
the completeness of the plan than we are about how the [venture]
founders reached their conclusions—what factors they considered relevant, what assumptions they made, how realistically they viewed
the competition, and so forth.*

A GE Capital banker concurs: "Everybody in this firm knows
how to make a spreadsheet sit up and beg. So it should come as no
surprise that our biggest arguments are almost always around our
different assumptions. . . . More often than not, our arguments are
about ideology. Some of us honestly
believe that a particular business or industry can only grow so fast or be priced so
high. The models reflect our assumptions.
When the models disagree, it usually
means our assumptions disagree."

**The biggest arguments
are almost always
around different
assumptions.**

Shubik notes that sophisticated spreadsheet models can reveal
logical flaws and financial inconsistencies in fundamental assumptions. But he freely acknowledges that dueling spreadsheets are
simply surrogates for competing, deeply held assumptions: "It's true
for spreadsheets," says Shubik, "and it was certainly true for the military simulations we surveyed. . . . Our real battles were over the
assumptions, not the models themselves."

The politics of spreadsheets

A budget negotiation between rival divisions of a fast-growing
global enterprise. An analysis of why a particular merger partner
makes the most strategic sense. A decision to vastly downsize an
organization. Spreadsheets play an important role in all three. But
the ultimate decision is likely to be more a product of organiza-

tional politics than of rational analysis. Simply put, spreadsheet models of important enterprise options are inherently political. The politics of spreadsheets matter as much as their methodologies. Indeed, spreadsheet methodologies themselves often *are* political.

In a 1996 *California Management Review* article, "The Politics of Forecasting: Managing the Truth," Craig Galbraith and Gregory Merrill detail the disturbing results of a wide-ranging survey of financial modelers in the high-level forecasting staffs of various domestic and international companies:

- 45 percent of the respondents reported that, after reviewing sales/revenue forecasts, senior management asks staff to adjust revenue projections to a more favorable level;

- 52 percent reported requests to adjust cost projections to a more favorable level;

- 42 percent reported requests for "backcasts"—that is, senior management predetermines an "appropriate" sales/revenue level and then requests numbers to support that level;

- 37 percent reported that senior management determines an "appropriate" future financial position, then requests pro-formas to support the decision;

- 26 percent reported that divisions/departments withhold useful data from other departments.

The authors observe that "misinforming the public, under the guise of forecasts and computer models, appears to have taken firm root in the culture of many organizations." But internal politics turns out to be the biggest impetus to model modification. Almost 60 percent of Galbraith and Merrill's respondents stated that their main rationale for adjusting forecasts and models was "to influence internal resource allocations." (Presumably the other 40 percent are liars.)

The issue here isn't whether spreadsheet politics is good or bad for the enterprise — or even whether spreadsheet politics is fundamentally honest or dishonest. The reality is that spreadsheet modeling has evolved into a medium for managerial Machiavellis. As spreadsheets have become more pervasive — as they have become a dominant medium for communication and collaboration — their political importance has also risen. Brand managers and divisional executives cannot afford to ignore whether their assumptions are built into enterprise forecasts. Conversely, as spreadsheets have become more political, they have also become more important. The main reason? Managers recognize that "winning" a spreadsheet battle over a contested assumption or a brilliant backcast sends a clear signal to the organization about the managerial pecking order. If Finance's spreadsheet models are consistently adopted in preference to, say, Manufacturing's or Marketing's, that says something about power and influence, not just intellectual acuity. It's become impossible to divorce the politics of spreadsheets from the politics of the enterprise.

> **Spreadsheet modeling has evolved into a medium for managerial Machiavellis.**

Prototypes that create new realities

Mark Twain observed that "if you've got a hammer, the whole world looks like a nail." A prototype, like a hammer, can also suggest a complete perspective on the world. Spreadsheet software has transformed perceptions of business as profoundly as the telescope transformed perception of the heavens. Spreadsheets aren't merely tools for seeing; they offer a way to create a new reality from a different point of view. As a result, spreadsheets can take on a life of their own.

For example, a former First Boston associate recalls frequent requests to perform backcasts to generate numbers justifying a particularly pricey merger or acquisition. Understandably, such financial finagling frequently yielded surreal numbers and inexplicable spreadsheet discontinuities. The resulting cash-flow kinks and investment twists turned out to be a source of intellectual stimulation. "We had to come up with perfectly plausible reasons why a mature market would suddenly grow by 20 percent," she recalls, "or why a particular asset would suddenly be worth 25 percent less." In other words, the artifacts of spreadsheet manipulation can inspire new ideas and stories to describe the deal. People were forced to use the numbers to discover more creative and ingenious ways to explain the fictional story they told. The nature of the numbers drove the deal as much as the nature of the deal drove the backcasts. The investment bank's smartest and most articulate minds reorganized their thoughts around the spreadsheet's model of proposed reality. These kinds of tales are typical. Models are used to tell stories; stories are used to explain models.

Spreadsheets create new realities by reframing old ones. At a major Midwestern bank, the new pricing manager was very concerned about its competitors' pricing. In fact, the bank's approach to pricing was simply to find out what other banks were charging. Using a very basic spreadsheet model, the new pricing manager analyzed the cost structure for the bank's direct and indirect loans. He plugged in the expenses, loss ratios, reserves, and expected life of the loans and determined a customized pricing model for the bank. His bosses had suspected that they would have to raise prices to meet profit targets. To their astonishment, the new manager showed them that their cost structure was among the lowest in the industry. The bank was particularly efficient in processing third-party loans.

With this new model, the executives' perception of reality changed. The bank lowered interest rates on most loans by an average of 1 percent and watched its profits soar. Indirect loans leapt from $300 million to $1 billion; direct loans doubled from $120 million to $250 million. Market tracking also became easier. When the Federal Reserve Board announced interest-rate changes, the bank could issue a new pricing sheet for its loans within one hour. In sum, the new pricing model helped create a self-fulfilling prophecy: By offering a credible model of reality, the spreadsheet became a mirror of reality. The perception became reality.

By offering a credible model of reality, the spreadsheet became a mirror of reality.

The imperative that models create their own realities is, however, fraught with risk, as the saying "Garbage in, garbage out" suggests. In financial-service firms, however, spreadsheets have consistently demonstrated the ability to inspire their own realities and build momentum for themselves. So long as the benefits are perceived as consistently and significantly outweighing the costs, the trillion-dollar-plus global finance industry will happily remain in their thrall.

These lessons all affirm that models can change minds, behaviors, and relationships both within and outside the organization. They also suggest two overarching themes that merit special attention.

1. Changes in a model lead to changes in interactions; changes in interactions lead to changes in the model.

Models and the human interactions and conversations that go on around them cannot be arbitrarily divorced. There is an ecological relationship between interaction and iteration: new interactions lead to new iterations, and new iterations lead in turn to new interactions. Models turn out to be more about mediating interactions between people than mediating interaction between information. They have a greater impact on interpersonal interactions than individual cognition. Steven Levy has observed

There is an ecological relationship between interaction and iteration.

that the simulation prowess of spreadsheets doesn't just create new information; it redraws organizational boundaries.

The what-if factor has not only changed the nature of jobs such as accounting, it has altered once rigid organizational structures. Junior analysts, without benefit of secretaries or support from data

processing departments, can work up fifty-page reports, complete
with graphs and charts, advocating a complicated course of action
for a client. And senior executives who take the time to learn how to
use spreadsheets are no longer forced to rely on their subordinates for
information.

The challenge for organizations is to productively manage both
the iterations of the model that change interactions and the interactions around the model that change iterations. That's how value
gets created and managed within the firm. Managing the economics of spreadsheet interaction and iteration has become a core competence in global finance.

2. Using spreadsheets as a modeling medium has proven less expensive—
economically, culturally, and organizationally—than not using spreadsheets.

There came a point when financial-service institutions literally
could not afford to ignore or underestimate the potential of spreadsheets. The cost of not using spreadsheets
overwhelmed the costs of using them. A
financial firm could no more dismiss
spreadsheet software than it could pull the
phone lines off the trading floor.

> **Managing spreadsheet
> iteration has become a
> core competence.**

There are two short stories that high-
light this overwhelming triumph of the spreadsheet. The first is
that, within five years of its creation, every student at the Harvard
Business School was required to use spreadsheet software. Dan
Bricklin's digital innovation had become the most rapidly adopted
standard in the history of American graduate education. The second story comes from a Lazard Freres managing director who
arrives at client meetings armed with deal books obese with spreadsheets representing hours of debate and exhaustive analysis. He
places these books on the table in front of the client. He says that
he knows he has truly been successful in managing his relationship
with his client if the client can be persuaded to do the deal without
ever once opening those books. "But," he notes with a tone of rue-
ful resignation, "they have to know the spreadsheets are there."

MODEL BEHAVIOR

Part II

3

OUR MODELS, OURSELVES

At Royal Dutch/Shell, the world's second-largest oil company, senior executives used to be urged to come up with three scenarios whenever they considered a strategic course of action. Each scenario was typically a small jewel of narrative analysis and foresight. But there was a catch. "The problem was we always chose the middle one," Shell UK head Chris Fay told the *Financial Times*. "So now we only put forward two."

Whether one finds it amusing or pathetic, this anecdote reaffirms a central thesis of this book: that organizations manage themselves by managing their prototypes. The deliberate decision not to model a middle path may yield greater insight into Royal Dutch/Shell's business culture than the scenarios themselves. Will debate be sharper without the option of a compromise choice? Shell management clearly thought that cutting the number of scenarios would hone the edge of its strategic conversations. What one chooses to model makes a difference.

The prototypes that organizations create reflect their perceptions of reality. Otherwise why build them? But they also reflect the organization's own internal assumptions about risk and reward. This is why prototypes always reflect the corporate cultures that create them. In the words of the distinguished University of Pennsylvania technology-systems historian Thomas Hughes, prototypes are "congealed culture": tangible slices not only of technology and technique but also of the corporation's own interpretation of market and cultural forces.

Whenever organizations attempt to create new product and service offerings at a profit, culture matters. This is not to say that the economics of innovation matter less—only that organizational imperatives powerfully shape the "innovation economy" of the enterprise. The Toyota culture embodied in a Lexus projects a different design sensibility than the Cadillac produced by General Motors. The technocultural idealism embodied in Apple's Macintosh is inherently different from the organizational imperatives that shaped Microsoft's Windows. The do-it-yourself investment ethos of a Charles Schwab demands a different innovation infrastructure than the advisory relationships championed by a Merrill Lynch. Indeed, corporate culture—not just manufacturing economics—is a prime reason why the sole of a new Nike isn't the sole of a new Adidas.

> **Prototypes are "congealed culture."**

Both anecdotally and empirically, it seems clear that cultural patterns powerfully shape prototyping processes. A prototype is the product more of a culture than of a process. Organizational change initiatives that don't explicitly address the internal cultures of prototyping and simulation are ignoring reality. This holds particularly true of organizations that say they want to innovate better, faster, and with greater impact on their markets.

To understand and influence how corporate cultures innovate, we need to better understand the roles that culture plays in creating new prototypes. A company's prototyping culture—the media, methods, and styles it uses to manage its multiple models of reality—offers a wealth of critical insights into how it designs and builds value. Acting on those cultural insights creates new opportunities for risk management and value creation.

World-class industrial designer-cum-engineers like IDEO's David Kelley and GVO's Michael Barry swear they can tell almost anything worth knowing about a company's innovation infrastructure merely by examining a few of its prototypes. "I could tell you absolutely everything," asserts Kelley, who runs the world's largest industrial-design firm, "from the care of the models to the quality of the thinking of the designers."

Why is that so important? "Because prototypes are a way of life," Kelley asserts. "I strongly believe that prototypes and products are intimately related, that the number of pro- totypes and quality of those prototypes [are] directly proportional to the ultimate quality of the product." The link between prototype and product, Kelley says, deter- mines meaningful innovation. This is not an idiosyncratic hypothesis. "In my experi- ence," observes GVO's Barry, "the companies that want to see the most models in the least time are the most design-sensitive; the companies that want that one perfect model are the least design-sensitive."

> **"The companies that want the most models in the least time are the most design-sensitive."**

Crude Mock-ups and Relentless Prototyping

When Sony president Nobuyuki Idei casually remarks that the consumer-electronics giant typically takes about a week to progress from a new product concept to a rough working prototype, the comment says less about Sony's technological prowess than about its predisposition toward rapid design. When 3M invokes its endur- ing motto "Make a little; sell a little" as inspiration for its intrapre- neurs, it is articulating a core value of relentless prototyping, not a bean counter's desire to analyze market demographics exhaustively. Sony and 3M are both global companies with undeniably strong prototyping cultures.

Creatively prolific organizations like 3M, Sony, and Walt Disney tend to generate lots of creative prototypes. By contrast, companies traditionally characterized by more rigid review structures and conformance to detailed specifications—Procter & Gamble, General Motors, and IBM throughout the 1980s, and AT&T and Kodak through the middle of the 1990s—tend to produce fewer, more elaborate and more expensive prototypes.

Quickly and continuously converting new product ideas into crude mock-ups and working models turns traditional perceptions of the innovation cycle inside out: instead of using the innovation process to come up with finished prototypes, the prototypes themselves drive the innovation process. To paraphrase MIT's Thomas Malone, "The more choices an enterprise has, the more its culture matters." The ongoing explosion of computational choice and the dizzying array of modeling options make culture—not just competence and cost—the critical success factor in managing innovation.

While the great ethnographies of simulation and prototyping cultures have yet to be written, fundamental differences in culture lead to qualitatively and quantitatively different products and services. This has been historically true and will become even more pronounced in the future. Understanding those differences and having the courage to act upon them is absolutely essential for any organization that wants to realign its culture with the coming opportunities. Managing organizational redesigns, process reengineerings, and core competences is not enough. Companies that want to build better products and services must learn how to build better prototypes and simulations.

Formal Processes and Informal Cultures

Most companies have formal prototyping processes and informal prototyping cultures, roughly comparable to their formal organizational charts and informal interpersonal networks. In some corporations, formal prototyping processes rule; in others, the informal prototyping culture is how the work really gets done. Royal Dutch/Shell has very formal processes for managing scenarios and simulations. The decision to reduce the number of scenarios to delimit managerial options is not casually made. At the same time, however, informal networks of lifelong Royal Dutch/Shell executives strongly influence how those formal scenarios will be interpreted and managed.

Hewlett-Packard's engineering-oriented culture assures that its engineering prototypes take precedence when new-product decisions are made. Yet HP also lets its new-product decisions be influenced by mavericks who have gone outside traditional channels to successfully demo prototypes to key customers. A creative tension between formal and informal management of prototypes exists. All innovative companies must manage the associated trade-offs between formal and informal culture. Either way, their prototypes will embody those trade-offs.

> **Most companies have formal prototyping processes and informal prototyping cultures.**

Redefining and reshaping the prototyping culture may prove to be the most important and provocative management challenge that innovative organizations now face. Organizations that want to understand their own prototyping cultures need to have the courage to confront what their prototypes say about their strengths and weaknesses. They must acknowledge where their models are more likely to create internal conflict and confusion than promote new awareness and consensus. Prototypes can be "wellsprings of knowledge," to quote Dorothy Leonard-Barton of the Harvard Business School, but they can also betray the whirlpools of organizational ignorance. Mapping those wellsprings and whirlpools is what all innovative organizations must do.

Prototyping Taboos: What's Missing and Why

The instinctive rational impulse when first examining an organization's prototyping culture is to carefully examine just what's being modeled. Yielding to that impulse is a mistake. Almost without exception, the most experienced and sophisticated students of organizational modeling look first for what is *not* being modeled. What is taboo? What does the organization absolutely refuse to model? What core assumptions are not permitted to play a part in a prototype design or simulation experience?

"I've learned that you learn far more about an organization from what they won't model than from what they do," asserts political scientist Garry Brewer, coauthor of the classic study of U.S. military simulations *The War Game.* "What I've observed—in both the military and private industry— is that organizations frequently leave out the very assumptions that are most important or most threatening to their sense of themselves. They always have a 'good reason' for this. . . . As a result, many organizations expend an extraordinary amount of effort developing models that can never be as useful or as valid as they say they want."

Always look first for what is NOT being modeled.

Unwillingness to question

Apple's first palm-top digital assistant, the Newton, failed miserably in the 1990s because of management's unwillingness to question the very marketing assumptions that got it mass-manufactured in the first place. Apple's leadership had insisted that Newton's value was contingent on its handwriting-recognition capabilities. The product was marketed around its handwriting interface. Without this ability to faithfully translate a user's penstrokes into legible text, the Newton would be computationally crippled. Unfortunately, Apple's handwriting-recognition software was never as reliable as promised. Reportedly, the software missed its performance milestones at every stage in the Newton's prototyping process. Apple executives never honestly confronted the taboo question of whether white-collar professionals would pay a premium for a handheld electronic organizer that could recognize their handwriting about 80 percent of the time. Indeed, no one had successfully challenged the fundamental design and marketing premise that handwriting recognition should be the organizing principle for a personal digital assistant. Honoring the taboo proved more important than honoring the customer. The Newton and its MessagePad successors ultimately became far better products, but they never overcame the public perception that their design taboos had created, and the product line was ultimately killed.

The PalmPilot, by contrast, was a runaway market success from the moment it was launched in early 1996, largely because it was the product of a far healthier prototyping culture. The PalmPilot's predecessor, the Zoomer, had also failed in the marketplace. But Palm's leaders, Jeff Hawkins and Donna Dubinsky, treated the failed Zoomer as a prototype for a next-generation handheld device. The result was a prototyping process aimed at building a different kind of product by creating a different kind of interaction with its target customers.

The PalmPilot was prototyped with a limited handwriting-recognition capability requiring use of a predetermined set of pen strokes. More important, the product was prototyped as a peripheral adjunct to its owner's personal computer, with which it could exchange data. PalmPilot designer Jeff Hawkins has emphasized how effectively the prototyping process challenged several initial interface-design assumptions. Bluntly put, the PalmPilot was an "anti-Zoomer," in the sense that taboos and unspoken assumptions were not permitted to derail successful design implementation. The fact that the PalmPilot was prototyped and engineered as a low-cost product, while the Newton was positioned as "far higher-priced but worth it," didn't hurt sales either.

Managerial arbitrariness

As Brewer and others observe, there are almost always "good reasons" for organizational taboos. Internal politics and personalities may matter far more than, say, legal considerations or obliviousness to market forces. Whatever the real motivation, honoring a taboo is a choice that an organization makes. The costs associated with observing those taboos, however, are frequently understated or ignored outright. Not even the most technically robust modeling methodology can survive arbitrary managerial taboos. Far more than technical fluency or organizational competence, this cultural imperative determines the potential utility of prototypes.

> **There are almost always "good reasons" for organizational taboos.**

An example from the *California Management Review* underscores the point: A consulting team presented to an international hotel firm a computerized site-location prototype designed to predict hotel occupancy at proposed business sites. Such a program could be a powerful tool for screening hundreds of building proposals from aspiring franchisees. Inputs included socioeconomic, demographic, and competitive factors, including "number of competitor hotel rooms in area." It was this last factor that raised the ire of the senior executive at the presentation. "I notice you have included [Hotel H] as a competitor," he observed. "This is a 'no-no'—the official attitude of this company is that we are more 'upscale' than [Hotel H]." The consulting team was required to include only four prescribed hotel chains, even though the explanatory power of the prototype was severely compromised.

Just as one fundamental criterion of a culture is its ability to declare individuals as insiders or outsiders, prototyping cultures are distinguished by their power to declare modeling assumptions either legitimate or off-limits. Prototyping culture begins with the decision—unspoken or explicit—of what will or what won't, what can or what can't, what should or what shouldn't be modeled. The power to proclaim any assumption sacrosanct or taboo has extraordinary implications for the management of any enterprise. But the implications for managers who need prototypes to manage expectations are especially profound.

> **The power to proclaim any hypothesis sacrosanct or taboo has extraordinary implications.**

Questions of culture

Nowhere are these implications more serious than in the richly modeled domain of national security. In its war games during the 1980s, for example, the U.S. Navy would not allow aircraft carriers—its biggest, most expensive, and perhaps most controversial weapons platform—to be sunk hypothetically. This taboo persisted even after the Argentines successfully sank a British carrier during the Falklands War. It held fast even when the navy's own submariners argued that carriers were particularly vulnerable to undersea attack. For a variety of budgetary, political, interservice-rivalry, and national-security reasons, the navy was permitted to run extensive war games and simulations in which its biggest and most vulnerable carriers were given a pass. The taboo was tacitly respected in virtually all formal reviews. External efforts to simulate conflicts in which carriers were destroyed were met with threats of security classification. One result, documented in Thomas B. Allen's *War Games*, a popular history of U.S. war gaming, is that the navy acquired a reputation for cheating that undermined the credibility of naval proposals and exacerbated interservice rivalries. This particular taboo was deeply ironic because, as Harvard's Stephen Peter Rosen ably documents in *Winning the Next War*, simulations and war games had been largely responsible for encouraging the navy to adopt aircraft carriers in the first place.

There is a difference, of course, between hypotheses that are taboo because they contradict cherished beliefs and those that are dismissed because they're considered wildly unlikely. Depending on an organization's risk culture, outlandishly improbable scenarios may be as much of a modeling taboo as the CEO's pet peeve. Resources are scarce; time is pressing. Organizations understandably don't want to devote much attention to low-probability events. But managers sometimes kid themselves about likelihoods. After all, the *Titanic* did sink. The *Challenger*

> **Organizations don't want to give much attention to low-probability events.**

space shuttle did explode. The Chernobyl nuclear plant did melt down. It's often difficult to determine whether an organization excludes potentially significant factors from consideration because they are culturally taboo or because their significance is underappreciated. But prototypes built on willful ignorance or taboos seldom survive prolonged contact with reality. Their costs consistently exceed their benefits.

The task of building a meaningful prototype forces an enterprise to either explicitly acknowledge its taboos or self-deceptively ignore them. This is why Brewer believes so strongly that unexplored assumptions can reveal more than those assumptions that are articulated. The real danger of taboos is that they fundamentally distort the internal marketplace of ideas and interactions. They breed dishonesty. Perhaps the most decisive step an organization can take to improve its prototyping culture is to track, log, and continually revisit the modeling assumptions it says it doesn't take seriously.

> **Taboos distort the internal marketplace of ideas and interactions.**

Blind spots

All models are attempts to manage complexity by making it simpler and more accessible. In the words of the noted economist Paul Krugman:

> *Any kind of a model of a complex system . . . amounts to pretty much the same kind of procedure. You make a set of clearly untrue simplifications to get the system down to something you can handle. . . . And the end result, if the model is a good one, is an improved insight into why the vastly more complex real system behaves the way it does. But there are also costs. The strategic omissions involved in building a model almost always involve throwing away some real information. . . . The result is that the very act of modeling has the effect of destroying knowledge as well as creating it. A successful model enhances our vision, but it also creates blind spots, at least at first.*

All prototyping cultures are cultures of deliberate simplification and oversimplification. But what should be simplified and why? At what moment does that useful simplification skid into dangerous oversimplification? At what point do these simplifications hit diminishing returns?

Simplification cannot be meaningfully discussed without reference to the medium it inhabits. Simplifying a 3-D model in computer software is different from simplifying a 3-D model made of clay. Is the model based on mathematical principles inaccessible to most managers? Is the prototype built in software that simulates two but not three dimensions of a physical product? Does the spreadsheet use corporate assumptions at the expense of the business unit closest to the customer? Which modeling media are most appropriate for effectively managing simplification?

Dueling Representations: Specs and Prototypes

Virtually all product and service innovations result from dueling representations: a "wish list" of specifications and the prototypes that attempt to embody them. Often our prototypes confirm that what we wish for is unrealistic or ill conceived. Conversely, they can also reveal that our wishes were not imaginative enough. Specs and prototypes can be mutually reinforcing, or they can prove implacable enemies.

> **Our prototypes can reveal that our wishes aren't imaginative enough.**

The tension between specs and prototypes is not unlike that between theory and experiment in physics; theory describes what's supposed to happen and experiment tells you what does happen. The culture of physics has always been a symbiotic battleground between theorists and experimentalists. Sometimes, theory has dictated the experimental agenda; at other times, experimental insights have driven the theoreticians. Similarly, managing the dialogue between specs and prototypes is essential to design innovation.

When the dialogue is poorly managed or breaks down, the results can be horrendous. It's not unusual for an organization to spend thousands of hours developing detailed specifications that are invalidated by the initial prototype. This pattern is particularly commonplace in software development, medical instrumentation, systems integration, and airplane cockpit design. Conversely, a common complaint about industrial designers is that their breathtaking prototypes prove impossible to manufacture cost-effectively.

Simply put, some innovation cultures are spec-driven while others are prototype-driven. Small entrepreneurial companies built around a brilliant product concept tend to be prototype-driven. Companies that need to coordinate large volumes of information while managing a large installed base of users—companies like IBM, AT&T, General Electric, and Aetna Life & Casualty—tend to be spec-driven. Prototype-driven cultures can evolve into spec-driven cultures and vice versa. Whether those transformations happen by accident or by design is also a question of culture.

There is always an ongoing relationship between iterations of spec-driven prototypes and refinements of prototype-driven specs, but it is usually obvious within the organization which mode of representation "calls the shots." Spec-driven cultures draw heavily from market-research data before concepts are moved into the prototyping cycle. In prototyping cultures, prototypes are typically used to elicit market feedback well before final versions of the product are tested. David Kelley of IDEO argues that innovation cultures need to "move from spec-driven prototypes to prototype-driven specs" to become more effective. That's arguable. However, few innovators argue that organizations prizing prototypes over specs—or vice versa—have fundamentally different design perceptions and processes.

> **It is usually obvious whether prototypes or specs call the shots.**

A clever and revealing prototyping experiment designed by software-engineering guru Barry Boehm reveals the sharp differences between these two approaches. Boehm had assigned seven teams of software developers to produce versions of the same cost-estimation model. Three used a prototyping approach; the other four were spec-driven. All produced roughly the same product with roughly comparable performance. But there were key quantitative and qualitative differences. The prototypers' programs used 40 percent less code and expended 45 percent less effort. The prototyped products rated lower in functionality and ability to handle erroneous inputs but significantly higher in ease of learning and use — that is, their customers liked them more. Particularly noteworthy, in Boehm's view, is that the prototyped software rated far higher in maintainability—contrary to the conventional wisdom that software that adheres to codified specs is easier to maintain than software produced by unplanned iterations of a quicker-and-dirtier prototype. In Boehm's experiment, the prototypers spent less time designing and programming and more time testing, reviewing, and fixing. The prototypes themselves clearly reflected the time spent responding to user feedback.

However, the prototyping approach had shortcomings. Prototypers spent less time documenting their efforts and produced less documentation than the specifiers. (It is plausible that, because users participated more actively in the prototype's development, comprehensive documentation was less necessary.) The prototypers' designs were also objectively rated as less coherent and more difficult to extend and expand than the spec-designed software.

Prototyping produced a smaller product of equivalent performance with less effort.

But Boehm's conclusions were unambiguous: prototyping produced a smaller product of equivalent performance with less effort. The productivity of the prototype-driven design, measured in user satisfaction per man-hour, was superior. The user interfaces were also rated superior, and what was psychologically significant was that there was always some part of the product that "worked." People like that.

Boehm also noted that conversations and interactions about the prototypes were different from those about the proposed specs. To call this a duel between prototyping culture and specs culture would be an overstatement. But it's readily apparent that two different design sensibilities led to fundamentally different design behaviors. Those behaviors in turn led to significantly different software products.

The Vocabularies of Prototyping

Same words, different meanings

The phrase *rapid prototyping* conjures radically different images in the minds of a software designer at a workstation and a mechanical engineer using stereolithography and laser sintering to produce physical three-dimensional models. The prototyping vocabulary dictates how people view the models they build. Design-intensive cultures tend to have richer prototyping vocabularies than companies that view prototypes as simply necessary stepping-stones to the final product. In some organizations, every product model built — no matter how rough — is called a "prototype." In cultures with strong technical and engineering traditions, however, only the working model that the organization has figured out how to manufacture is called "the prototype" — everything else is called a breadboard or a mock-up. "Prototype" assumes a special connotation.

Interdepartmental misunderstandings arise when engineering's "mock-up" is marketing's "prototype." One division's "roughs" are another's "comps." Does the organization have alpha, beta, and gamma prototypes, or do versions get labeled by number? Are prototypes "designed" or "built"? Do people "create" prototypes or "assemble" them? Are prototypes regarded internally as engineering artifacts or as marketing samples? Is different language used to describe internal prototypes and those shown to customers? When the same words mean different things to different people, misunderstandings are inevitable. Different computer-aided design and computer-aided engineering (CAD/CAE) software like CATIA and

Autodesk AutoCAD also generate their own argot and idioms. Indeed, some engineers are more comfortable calling a three-dimensional CAD virtual simulation a "prototype" than its foamware counterpart.

The prototyping medium itself also shapes language. Mention "foamware" to some design teams, and they'll enthusiastically recall how they carved up multiple approaches to a new product. At product companies that don't make foam models, you'd get nothing but a blank stare.

Perhaps the most profound impact that prototyping has had on language is the new vocabulary it can generate for new features. IDEO's Kelley recalls designing a series of toothpaste-tube prototypes for Procter & Gamble. One criterion for assessing these prototype tubes was their "suckback": how much toothpaste the tube sucks back after the user stops squeezing it. Prototyping enabled designer and client to create a working vocabulary to calibrate the tube to the desired level of suckback.

> One criterion for addressing prototype toothpaste tubes was "suckback."

Different prototyping cultures develop their own languages and vocabularies. Understanding a given prototyping culture without learning its language is like trying to understand accounting without using numbers.

Differences in fluency

A congressman who favored a "soft-technology" approach to U.S. energy needs was discussing demand projection with a modeler. He pointed out that adequate conservation measures and modest lifestyles could reduce growth of electrical demand to 2 percent per year. "But Congressman," said the modeler, "even at 2 percent per year, electrical demand will double in thirty-five years."

"That's your opinion!" exclaimed the congressman.

The point is not to mock legislative innumeracy but to underscore that even the most technically valid model doesn't guarantee useful communication or productive interaction. Different organizations have different languages. Finance-trained MBAs are typically far more comfortable manipulating spreadsheet simulations than are liberal-arts-trained salespeople. Chemical engineers may model complex processes as arrays of differential equations that baffle the technicians who run the plant. Industrial designers might render new products as two-dimensional sketches that manufacturing engineers consistently have trouble visualizing as three-dimensional objects. What kinds of conversations help bridge those gaps? Who manages them?

We use models to tell stories, and we use stories to build models. As management theorists and practitioners from Peter Drucker to Karl Weick to GE's Jack Welch agree, the stories that organizations tell themselves are at the heart of how meaning gets created. So too, then, are the models that arise from those stories.

Prototypologies

"Prototypes are designed to answer questions," David Kelley asserts. Different questions require different kinds of prototyping media, such as foamware, stereolithography, or computer simulation. Is a given question best addressed by "high-fidelity" prototypes of the product or service? Or are "low-fidelity" quick-and-dirty simulations the best way to go? Different prototyping cultures employ different prototypologies—types of prototypes—for answering such questions.

Is a given question best addressed by high-fidelity prototypes?

It's important to note that sophisticated questions do not necessarily call for sophisticated prototypes or simulations. Similarly, seemingly simple questions may defy even the most creative prototyping efforts.

Typically, prototypes and simulations are constructed to answer questions about the following:

- speed
- manufacturability
- cost
- feasibility
- functionality
- new materials

- reliability
- training
- appearance
- usability/user interface
- maintenance

Prototypes and simulations can do more than answer questions; they can also raise questions that had never been asked before. Playing with a prototype can stimulate innovative questions as surely as it can suggest innovative answers. The best and most powerful models are provocative, and the unexpected questions that a model raises are sometimes far more important than the explicit questions it was designed to answer. There are profound cultural differences between organizations that build prototypes primarily to create questions and those that do so to answer questions. The ratio of questions asked to questions created says a lot about the organization's innovation culture.

Some organizations want prototypes to create questions; others want them to answer questions.

Prototype portfolios

The mix of prototypes—and the development budgets associated with them—represents the organization's "prototype portfolio." Just as an investment portfolio embodies the risk/reward equation that expresses the investor's expectations, the prototype portfolio yields insight into the organization's own innovation/market expectations. There are prototyping bulls and bears, cash cows and question marks, stars and dogs. Prototypes are as much media for managing risks as for exploring opportunities. They can be treated as an insurance policy or as an option on the future.

Whether they explicitly acknowledge it or not, all companies that build prototypes are prototype-portfolio managers. Issues like return on investment, reinvestment, and divestiture are all strategic innovation issues as much as strategic financial issues. They are also issues of strategic culture.

Similarly, companies that focus their creativity and budgets on "looks-like" prototypes have wildly different priorities from firms that invest heavily in "works-like" prototyping. "Engineers generally ask what the product *does*," says Carnegie Mellon design professor Dan Droz. "Many industrial designers ask 'What is the product?'"

Thomas Alva Edison's Menlo Park workshop was dominated by functional "works-like" prototypes that embodied its machine-shop culture. Xerox Corporation—once obsessed with prototyping for manufacturability—now devotes considerable resources to interface/usability prototyping initiatives. Chrysler and Toyota have both become far more aggressive in prototyping the potential of new materials in their automobiles.

Some prototypes are presented as design platforms to elicit feedback; others are sales tools designed to procure additional funding. Like companies that keep two sets of books, some organizations keep two sets of prototypes: one for show, the other for work. IBM developers used to be notorious for showing one kind of prototype to top managers and another to fellow technologists.

Apolitical prototypes probably answer unimportant questions.

Political questions, however, are just as valid as any other organizational issue. Indeed, apolitical prototypes are probably answering questions that the organization doesn't consider important. Prototypes and simulations are always political. If one accepts the aphorism "knowledge is power," managing prototypes and simulations is about managing power and influence. The organizational conversation shifts from the power of culture to the culture of power.

As we have seen, some prototypes raise political questions that the organization is unwilling or unable to answer. A primary reason for the failure of the IBM PCjr home computer in the mid-1980s was that IBM management had decided it might cannibalize sales from IBM's popular line of personal computers. The product of a spec-driven culture, the PCjr was deliberately hobbled in the prototyping process to thwart that possibility. Less than two years after its introduction, the PCjr was withdrawn. IBM's internal politics of prototyping killed it.

Ultimately, one of management's greatest challenges is to effectively integrate its diversified portfolio of prototypes into an integrated product or family of products. That is the arena where politics, economics, and organizational culture are often in sharpest conflict.

Prototyping media

Can Detroit's lagging competitiveness in the 1980s be blamed in part on its prototyping media? Absolutely. Intricate and expensive clay models didn't lend themselves to easy modification or rapid iteration. The sheer effort required to craft them actually made them more like untouchable works of art than malleable platforms for creative interaction. The medium's message is, Look but don't touch. "American automobile companies didn't have an iterative culture," says IDEO's David Kelley. "Clay . . . was like God's tablets." GVO's Michael Barry agrees: "When a model starts to harden up," he says, "so does a lot of the thinking." Clay was more than a medium; it was a metaphor for management.

> "When a model starts to harden up, so does a lot of the thinking."

Daniel Whitney of MIT's Draper Labs, who has studied the use of computer-aided design tools in Japan, observes that until the 1990s, U.S. car companies attempted to use clay models as inputs for their computer-aided design systems. This approach combined the worst of both media worlds: it was labor-intensive and imprecise, analogous to typing a handwritten novel into a word processor, editing the printout by hand, and retyping the final version into a computerized typesetting system. The cost in time, labor, and errors is painfully high.

Toyota took the opposite tack, insisting that its stylists use CAD tools from the outset. The clay model thus became the ultimate output of the CAD system. This policy utterly transformed both Toyota's prototyping process and its design culture. "Both Toyota and Nissan . . . studied the stylists and worked hard to come up with user interfaces and program features that reproduced the prior working environment and aesthetic skills and 'tricks,'" says Whitney. "Toyota even simulates on the computer screen the effect of strip fluorescent lights reflecting off the car's surface. So the stylist can judge the smoothness by 'eye,' just the way he used to in the clay studio, rather than by some arcane calculation he does not understand and will never trust or use."

Toyota was able to transform a new idea into a one-quarter-scale clay model in under thirty days. To no one's surprise, Toyota's speed-to-market proved twice as fast as General Motors' throughout the 1980s and early 1990s. More iterations in less time at less cost. The much-vaunted Toyota Production System's success was amplified by the automaker's strategic embrace of a new prototyping medium. Toyota's effective use of prototyping media helped make it a far more competitive automobile manufacturer. Design-sensitive companies like Chrysler took note.

It would be foolish to blame General Motors' decline on clay, but equally foolish to ignore the role of prototyping media in determining the speed and quality at which automobiles are built. "As the model-making material changes, the design process changes," says GVO's Barry. "Hewlett-Packard used to do its calculator modeling in cardboard, which explained why all their calculators featured angles and edges. When the company switched to foam, you saw calculators that had a softer and more rounded look. . . . Foam lends itself to certain kinds of subtleties." Barry and other designers argue that one of the biggest changes in industrial model-making, comparable to computer-aided design, is a material called Ren Shape—a high-density foam that sculpts well, holds a shape, but can't be dented by a nail. The

> "As the model-making material changes, the design process changes."

properties of this foam have given designers and their clients a far better way to render the look and feel of a product. Product teams that adopt Ren Shape design differently, Barry says. The new medium evokes new designs.

Ideological battles are beginning to emerge between the champions of "virtual" prototyping and "palpable" prototyping. Some cultures will expertly blend both approaches, but the rivalry still promises to dominate new-product design over the next generation.

> **Ideological battles are raging between "virtual" and "palpable" prototyping.**

Changing the medium

When a new prototyping or simulation medium confuses corporate discourse, management may reject it. Arie de Geus of Royal Dutch/Shell observed that "the computer itself was getting in the way of our primary purpose: understanding the system" by demanding a series of new behaviors from senior management. "We were asking too much," he recalls. "So we eliminated the computer. Instead, we moved to a distinctly low-tech technique: noting our ideas on magnetic colored hexagons, which are placed on a whiteboard so that everyone can see what is written on them. We then cluster and rearrange the hexagons at will to show related concepts or connections between ideas."

Prototyping is inherently a multimedia process. The prototyping media that organizations choose—or choose to ignore—are deeply revealing of how, for example, a products culture creates value. Just as revealing is the interplay between these disparate media. Most product prototypes begin in two-dimensional form, as sketches or software-generated imagery. Many companies specify standard formats or genres for their drawings. Sony, for example, insists that its consumer-electronics designers rigorously conform to an exploded-view format, while 3M's sketches are considerably less formal.

> **Prototyping is inherently a multimedia process.**

In the early 1990s, Timex dramatically changed its product-development process by adopting a new two-dimensional prototyping design medium. According to John Houlihan, the company's director of industrial design, Timex replaced its fifteen-person mock-up source with ALIAS CAD software, which creates photorealistic renditions of Timex watches. The company shows these "photo" models both internally and to customers. Houlihan admitted at the time that "the sales and marketing departments are skeptical. They are used to having mock-ups for sales presentations. They don't embrace my notion regarding the importance of photorealistic images." Changing the prototyping medium challenged the prevailing culture at Timex. Houlihan acknowledged that the leap from photorealistic watch to finished watch became a serious management challenge.

Managing the media transformation from two-dimensional representation to three-dimensional model also becomes a culture-defined process. "In some cultures," says GVO's Barry, "The 2-D is malleable but the 3-D is not." One of GVO's clients wanted to postpone turning product drawings into an actual physical model out of fear that the design dialogue would end once the hard prototype appeared. "They thought the design would freeze with the first model," says Barry. The company's new-product culture made the transition from one medium to another extraordinarily difficult. In immature design cultures, hard models can harden thinking.

Barry and other leading industrial designers emphasize the importance of transitional media like "rough" foamware to help bridge the dimensional gap. The value of foamware lies in its very roughness and lack of color. "The minute you lay in color," Michael Barry observes, "you finalize it. . . . You send a cue that it's 'finished.'"

Most companies, says Barry, have been two-dimensionally intensive: sketches and drawings proliferate, but they build only a couple of hard models and virtually no transition foamware. He observes that perhaps one-third build a few transition foams and one hard prototype. Fewer than 10 percent actually build more than ten or fifteen foamware models.

Dan Droz of Carnegie Mellon points out that "roughness" too is culturally defined. Some organizations are very comfortable with rough prototyping media, whereas others push toward perfection as soon as possible. The strength of rough prototyping media is that they encourage playing with ideas, possibilities, and potential. Roughness encourages questions. "Too many organizations believe that manageability means predictability," says Droz. "The idea that you can 'play' your way to a new product is anathema to managers educated to believe that predictability and control are essential to new product development."

> **Roughness encourages questions.**

Cost is a constraining property of prototyping media: it is far easier to experiment with inexpensive media like paper, cardboard, and foam than with a harder medium like clay or steel. Simple economics dictate that organizations using inexpensive prototyping media tend to generate more prototypes per unit of time than organizations that use more expensive media.

As rapid prototyping media like stereolithography and laser sintering decline in cost and complexity, physical prototyping processes undergo a transformation as profound as what happened when electronic media eclipsed print media. What desktop publishing did to graphic design, desktop prototyping will do to new-product design and desktop simulation will do to process and interaction design.

But do not ignore the provocative question that Michael Barry poses about the core management issue a prototype or simulation can raise: "Is it real enough to give power?" If the modeling medium doesn't reflect and respect the questions of power, it may end up offering insights without influence.

Process maps

Michael Hammer and James Champy, coauthors of the best-selling *Reengineering the Corporation*, have observed that the most valuable part of a corporate reengineering engagement often occurred when employees mapped their existing processes. Spaghetti-like workflow diagrams, thumbtacked to conference-room walls, often represented the first time the organization could see how its work processes really flowed. Some software-sophisticated organizations tried to model their processes digitally. These process maps contradicted the formal "org charts" that purportedly depicted how the about-to-be-reengineered organization operated. The ensuing conversations around those maps began a process of cultural, not just organizational, reengineering.

But the key point is that an organization can't meaningfully reengineer around a spreadsheet; it's the wrong medium. Processes could only be reengineered after they had been visually modeled or simulated in a way that made sense to the enterprise itself. Bluntly put, reengineering would have failed even as a management fad if it had not offered a persuasive prototyping medium as a precursor to its implementation. Arguably, process mapping was the single most important element of reengineering's conquest of the Fortune 1000. A review of the reengineering literature confirms this. The most effective reengineering was contingent on a prototyping culture—a culture ready and able (if not particularly willing) to map the processes it was reengineering. This principle is equally applicable to organizations committed to transforming their prototyping cultures.

> Reengineering would have failed without a persuasive prototyping medium.

Time and Cycles

As anthropologist Edward Hall convincingly argues in *The Dance of Life*, different cultures perceive time differently. The same thing is true of prototyping cultures. The role of time is one of the clearest markers with which to distinguish cultural priorities. And in an era when conventional managerial wisdom holds that speed-to-market is the key to competitiveness, the time dimension assumes even greater significance.

How can an organization champion speed as a competitive weapon without transforming the speed at which it designs, builds, and tests prototypes? Remember, Timex's motivation for switching to computer-driven photorealism was to accelerate its product-development cycle. How organizations choose to manage time strongly influences how they manage prototypes. For example, Sony and 3M take great pride in the speed with which they can produce a functional prototype. Companies like Microsoft and LSI Logic, a custom silicon-chip designer, also boast exceptionally quick mean-times-to-prototype.

Other companies—notably IBM, AT&T, General Motors, and Johnson & Johnson—have characteristically studied a design concept for weeks before even beginning to turn it into a prototype. Their mean-time-to-prototype is often months slower. The culture of such companies treats the prototype as an end product of thought, not as a partner in the thinking process. Companies with fast mean-times-to-prototype usually generate more prototypes and run through more cycles than those that move more slowly.

> **It is hard to persuade companies that one more iteration costs less than a flawed product.**

As a rule, the more prototypes and prototyping cycles per unit of time, the more technically polished the final product. Most companies, however, have resolved on a fixed number of prototyping cycles. This number is typically a function of tradition and culture as much as economic or competitive necessity. Only a handful of companies complete as many as five prototyping iterations in a year. However, IDEO's David Kelley says it is consistently difficult to persuade companies that performing one more iteration is far less costly than releasing a flawed or incomplete product. "The obstacle is tradition and cost," he notes. "Most companies really don't know how to manage the evolution of their prototypes over time."

Like writers who believe that one or two drafts are adequate, many design teams consider that one or two iterations of a "works-like" prototype—"to get the bugs out"—sufficient. In a perversion of Parkinson's Law that work expands to fill the time available, new-product-design teams frequently spread out iterations so that three prototyping cycles result whether the time available is six months or eighteen months. (With longer spans of time, each prototype does tend to be more fully developed.)

But growing emphasis on time is radically redefining prototyping culture by introducing a practice that Harvard's Steven Wheelwright and Kim Clark call "periodic prototyping": instead of producing prototypes as they deem appropriate, design teams are required to produce them on a fast fixed schedule. At Honda, for example, an automobile might go through ten or twelve trial builds, requiring a new prototype every two weeks. Motorola instituted a similar schedule for its popular Bandit pagers.

In other words, many time-sensitive organizations are now institutionalizing the prototyping process—and thus redesigning the culture—around explicit deadlines and schedules. The culture is now force-marched to double or even triple the number of prototyping cycles per unit of time. A familiar theme reemerges: quantitative changes inevitably lead to qualitative changes. "When you are producing more prototypes over less time," says Carnegie Mellon's Droz, "you talk about them differently and treat them differently. You become more concerned with integration questions." That is central.

> **"When you produce more prototypes over less time, you talk about them differently."**

In *Technology Integration,* Marco Iansiti of Harvard Business School emphasizes the importance of rapid mean-times-to-prototype and multiple iterations and singles out three fast-moving Internet companies—Yahoo!, NetDynamics, and Netscape—as examples of different approaches to rapid prototyping:

> *Although early prototyping is common to all firms, the sophistication of early prototypes varies considerably. At Yahoo!, a demonstra-*

tion version of My Yahoo! was produced one month into the six-month development cycle and consisted only of a mock-up of the HTML pages that would form the user environment for the service. At NetDynamics, a mock-up of the user development environment was also produced and shown to customers during the first month of the project to inform decisions about desired features. At Netscape, however, the first integrated prototype version of 3.0 . . . was relatively comprehensive. Many new functions had been integrated, representing over 50 percent of the new code required.

Iansiti points out that all three approaches serve a common design ethic. "As the pulse of the project," he notes, "prototypes punctuate the effort by integrating architecture and technology into a coherent package." Iansiti asserts that experience is the key to successful integration management, and quotes a Yahoo! engineering vice president: "Experience is essential. It is the only thing that lets you see how the whole system works. It is critical to make the right trade-offs as the project evolves."

With all due respect to Iansiti and Yahoo!, when is it *not* important to make the right trade-offs? What is so provocative and important about the integration challenge that Iansiti describes is not just the trade-offs that experienced software engineers make but also the role that customers play in prioritizing trade-offs. In the rapid-prototyping, multiple-iteration environment of Internet product development, integration and coherence depend as much on savvy users as on experienced developers.

Never Show Fools Unfinished Work

The single most important element of prototyping culture is, without question, who gets to be part of it and why. Absolutely nothing says more about an organization's modeling culture than its networks of power and communities of influence.

Who owns the prototype? Who manages it? Who gets to see it, and when? Who determines which constituencies have a say in the next prototyping cycle? Is there an internal "model shop" responsible for prototyping on demand? These are the questions that reveal most incisively the profile and pathologies of the corporate prototyping culture.

Perhaps the first fundamental question is, Who builds it? There is a world of difference between organizations with a "do-it-yourself/build-it-yourself" culture and those with model shops and analysts who build models for management. Systems analysis gurus C. West Churchman and Russell Ackoff have argued convincingly that a prime reason for the checkered history of "operations research" in American management is that few managers could build and use the statistical models on their own. Do the people who use the model help build the model? Or do they only manipulate a prototype handed to them by "experts"? Royal Dutch/Shell's de Geus explicitly rejects the latter practice: "That temptation should be resisted. Any model emerging from the modeler's back-room office runs an unacceptably high risk of being rejected by the [management] team. The modeler simply doesn't know what they know; the modeler can't make the model represent their understanding of their reality." Powerful words. They reflect a culture that places a premium on active participation. But who gets to participate in prototyping, and when?

> Perhaps the first fundamental question is, Who builds it?

At one highly regarded Silicon Valley company with a strong engineering culture, people are happy to show their bench prototypes—until the audience reaches the vice-presidential level. Then the unspoken but widely understood edict "never show fools unfinished work" kicks in. Top management finds it too difficult to see beyond prototype roughness to the ultimate product, and good ideas are often rejected for what is perceived as inadequate execution of the prototype. As a result, many engineers conceal their more provocative prototypes from senior management until they have been appropriately polished. "If you show them the plywood first," says IDEO's Kelley, "they can't make the conceptual jump. . . . The higher up in the company you go, the harder time they seem to have with visualization." Consequently, demonstrations to senior management assume the quality of theater performances rather than interactive dialogues. As a result, top-management feedback is solicited relatively late in the design cycle.

This is hardly an atypical situation. At many large organizations, demoing a prototype to senior management assumes all the trappings and investment of a Broadway musical. The prototype becomes a medium for proving a point rather than a vehicle to evoke discussion.

Any organization that wants a better understanding of its prototyping culture ought to measure the interval between initial creation of a prototype and the first demo for senior management. The longer the interval, the more likely it is that top management is being asked to approve—rather than participate in or react to—new-product creation.

As we saw earlier in the IBM PCjr episode, prototypes can become heavy artillery in departmental turf wars. "In many organizations," says GVO's Barry, "whoever owns the hard model wins." At packaged-goods companies, the brand manager typically owns the prototype. At high-tech companies, by contrast, engineers and technicians have traditionally owned the prototype and bickered with marketing and manufacturing over suggested modifications. More significantly, the modeling shop has typically reported to engineering and has been dependent on its budget. Want to swiftly change the prototyping culture of an industrial-products company? Take the model-shop budget away from manufacturing or engineering and give it to the marketing department.

In departmental turf wars, whoever owns the hard model wins.

As companies push toward cross-functional development teams, however, the role of the prototype frequently changes. As Dorothy Leonard-Barton of Harvard has observed, "As the community of stakeholders in the development of new products and processes broadens, physical objects that help bridge disciplinary and functional boundaries become more important." In essence, prototyping becomes a medium not just for interdepartmental integration but also for organizational redesign. Prototype-driven innovation ends up promoting a radical deconstruction of the existing organizational chart because it becomes increasingly important to avoid the departmental turf wars that cripple prototypes.

Zucchiniware

Even the most well-entrenched prototyping cultures can be dramatically transformed by the smallest of changes. From its beginnings in Albuquerque in the 1970s, Microsoft has had a very strong prototyping culture. But former Microsoft manager Julie Bick tells a story—which the company independently confirms—about how the world's most profitable software company made its prototyping culture even stronger and more inclusive.

Even entrenched prototyping cultures can be transformed by small changes.

One of the dullest low-level tasks in creating software at Microsoft is managing "the daily build," which is, in practice, a daily prototype of the product in process. The person performing the daily build collects all the code from the programmers on the product team and puts it on a single computer to see if it all works together. For years, this task was performed by an entry-level person and regarded as mind-numbing grunt work. One manager changed that in a way that made the process more efficient and more effective. Instead of delegating the task to a grunt, the manager gave the daily-build responsibilities to the people writing the code. Each day the programmers would give their code to one "buildmeister," who put it all together. If the code wasn't compatible, the person whose software "broke the build" became buildmeister as punishment until someone else's code broke the build. In the summer of 1996, the buildmeister was also given an enormous zucchini—"the zucchini of questionable freshness,"—sometimes with Groucho Marx glasses and a fake nose, to keep until the next buildmeister was named.

Delegating the task of buildmeister to the team changed Microsoft's daily prototyping process for the better. More developers got to see how their work fit together, or didn't. No one wanted to be buildmeister, so an extra incentive to hand in quality code was created. What's more, the unpleasant task of build management was equitably shared by everyone in the group. Accountability, responsibility, and quality were thus aligned.

This realignment had other important repercussions. The smartest and savviest high-level software developers hated being buildmeisters and wanted to spend as little time on the task as possible. But instead of weaseling out, they wrote tools to automate the task of buildmeister. The result? Microsoft developers now manage the build with a fraction of the friction and in a fraction of the time they did in the mid-1990s. In an environment of shared responsibility, they have become more efficient and more effective.

Might these improvements have happened anyway if the buildmeister role had remained in entry-level hands? Possibly. But what would have been the incentives to make it happen? Most of Microsoft's development groups recognized that this simple change promoted a culture of participatory prototyping that generated benefits far in excess of its costs and adopted it themselves. That is the essence of "best-practices" management.

Rehearsalware

Changing people's roles is an excellent way to change the culture. But changing the prototyping medium can also change roles. At VeriFone, a Hewlett-Packard transactions-technology subsidiary, changing the prototyping medium created new expectations and cultural challenges.

"Our design centers all reside in distinctly different parts of the world, and each center has its own culture," explains Brian Drugge, a VeriFone industrial-design manager. "So there can be a culture clash when projects involve personnel from multiple sites. You cannot take a decision-making process that works in California and expect it to be effective in Taiwan." Mechanical engineering manager Roger Hall agrees: "One example of a cultural barrier we have had to overcome happened when all our design data went on-line and became available for everyone involved in the project to see. Unfortunately, this goes counter to the culture in our Taiwanese facility. The mindset of most Chinese engineers is they don't want anyone to see what they are doing until they are done. We had to delay giving full access to design data in the past because of this hesitancy."

Changing people's roles is an excellent way to change the culture.

Mapping the flow of its models and prototypes would be one of the most revealing exercises an enterprise could undertake. Who gets to see what, and when? When are modifications made? Who requests them? Which requested modifications are ignored? This map, not an organizational chart, would probably be the best starting point to evaluate core-process redesign. Such a map would reveal how essential or marginal the formal prototyping process truly is.

Because a prototype is an artifact, it can be meaningfully tracked and measured. An organization can quickly discover the power points and political bottlenecks that govern the value-creation process. Metrics must become as integral a part of prototyping culture as accounting is for financial culture.

When it comes to creating a "best-practices" prototyping culture, however, the rules of the game appear to be shifting. To succeed today, the prototype can't be seen as the property of the engineers, designers, or marketers—it has to be seen and treated as community property. And the most important member of that community must be the customer. The customer must have an opportunity to play with and relate to prototypes as they evolve.

Organizations must ask themselves, When do customers and suppliers get to see the prototypes? Do customers and suppliers participate in the prototyping process? At VeriFone, according to a top designer, "vendors are on the team and must buy into a design.

Otherwise we don't do it." In the software industry, not one piece of personal-computer software has been successfully launched without input from potential customers about alpha- and/or beta-version prototypes. In the astonishingly capital-intensive micro-processor industry, Intel and Microsoft have typically worked closely throughout the entire prototyping cycle to reconcile operating-systems design issues with silicon-logic design constraints.

At some companies, showing outsiders the prototypes is no big deal. At others, only senior management can authorize such demonstrations. Both extremes have their problems: familiarity breeds contempt, and isolation breeds a noxious hybrid of arrogance and ignorance.

Not one successful piece of software has been launched without customer input.

However, there seems to be little doubt that collaborative prototyping with customers and suppliers can yield competitive benefits. IBM attributed much of the success of its old AS/400 minicomputer to participation by key customers. Microsoft's collaboration with applications-software companies was integral to the widespread acceptance of Windows. Boeing's networked interactions with its vendors enabled the 777 to meet its schedule and performance expectations. Nike's successful origins stem from intense collaboration between athlete and designer.

The essential question that emerges for tomorrow's organizations is, How do we best balance the community that creates the prototype with the communities that the prototype creates? Effectively striking that balance will not only display the organization's cultural priorities but also define the organization's ability to effectively innovate in an increasingly competitive marketplace.

Different organizations have fundamentally different notions about what kind of prototyping behaviors will work best for them. However, most organizations don't seem to be particularly aware that their innovations are governed by their prototyping and simulation cultures. Strategic introspection is lacking. Most organizations simply don't bother tracking how well or how often their people interact around iterations of their prototypes.

"Whoever discovered water," observed the late advertising sage Howard Gossage, "you can be sure it wasn't a fish." Organizations are often the last to recognize the cultural imperatives that guide them. Angela Dumas of London Business School uses the telling phrase "silent designers" to describe all those whose decisions or passive assent shape a new product, service, or offering.

When working with organizations, I frequently ask executives and project managers: If you had access to "rehearsalware" guaranteed to improve the quality of whatever you chose to rehearse—a presentation, a design meeting, a sales call, a brainstorming session—by 50 percent, for free, how much time would you and your organization spend using it? About one-third say they would practice significantly more, another third somewhat more, and about one-quarter the same amount of time, on the grounds that they would get more value for the time. The rest say they would practice less. In some organizations, everybody insists that they would religiously use the rehearsalware. The others say they would practice much less, because they're already happy with how much they practice and the rehearsalware would simply make them more efficient.

"Whoever discovered water, you can be sure it wasn't a fish."

In other words, there is absolutely no consensus about how people would use a technology guaranteed to improve performance. Individuals and their organizations have radically different perceptions of the value of preparation and rehearsal. But their intuitions are all context-free. Reliable answers to questions of this kind begin with measuring the processes and practices of prototyping.

4

P R O D U C T I V E   W A S T E

In 1993, Chrysler engineers took roughly one hundred days to attach the body of the first physical prototype LH vehicle to the chassis pallet. This pallet contained the engine, suspension, fuel system, fuel line, and a host of other vital components. Five years later the body and pallet attached conflict-free within five minutes. That represents a 28,000-fold improvement. Boeing enjoyed comparable successes physically assembling its digitally preassembled 777 jet. Both Chrysler and Boeing relied on CATIA design-and-engineering software.

A process that took one hundred days in 1993 took five minutes five years later.

For the LH program, Chrysler engineers stored some 1,900 drawings in their computers, represented by 11 million digital polygons. They could check for interferences caused by adding new parts in a fraction of the time previously needed. In one case, the computer performed 8,646 checks for interferences between sheet-metal components in seventeen seconds. Using the best paper engineering drawings possible, Chrysler's best engineers couldn't hope to do a comparably reliable check in seventeen days.

Time savings and money savings interacted dynamically to create a virtuous cycle. Digitalization of design allowed for prototype performance to be simulated cost-effectively in virtual environments. Automobile manufacturers committed to design software that could be integrated into engineering software that could be integrated into manufacturing software that could be integrated into digitally simulated environments. Seamless simulation became the goal.

Engineers can modify virtual vehicles before crashing them into real walls.

The ability to seamlessly port computer-aided designs into software ranging from finite-element analysis to computational fluid dynamics has dramatically reduced the perceived need for building physical prototypes. High-speed crash tests and fender benders can now be performed on-screen, allowing engineers to modify their virtual vehicles before crashing them into real walls. "We could get a feel for how the physics inside the computer related to the physics of the real world," says one Ford mechanical engineer. "We would make bets about how close the real crashes would be to the simulated ones. We got very good."

What finite-element analysis was to collisions, computational fluid dynamics was to automotive aerodynamics. CFD simulations let engineers optimize vehicle shapes and profiles prior to physical wind-tunnel testing. "You may go through a thousand iterations in the computer by the time you finally build your first prototype," the University of Michigan's David Cole told *Design News*. "The whole idea is to trade off computer time for shop time. That way, you're a lot closer to a finished product than you would be if you started by drawing lines on paper." Detroit's digital infrastructures have already had a bigger impact on design, manufacturing, and profitability in the automobile industry than Henry Ford's assembly line.

Digital Abundance

The ability to squander gigabits and terabytes on processors and networks now drives innovation culture. Computational wealth, not material scarcity, dominates design debates. A new era of technical and business trade-offs has emerged. These trade-offs—which militate for a degree of creative collaboration previously unknown to global enterprise—are fast becoming how firms define and differentiate themselves in the minds of their best customers and most critical suppliers.

Speed as a commodity

In industry after industry, in market after global market, the rise of digital modeling media has compressed development-cycle times and radically reduced costs. The numbers don't lie. The benefits are clear. But speed for the sake of speed is valueless. That's not business; it's self-indulgence. The fundamental dynamics have been seriously misunderstood. Management mantras to the contrary, speed and cost reduction are not the ultimate outcomes of the adoption of digital media; they are just intermediate stages.

Speed for the sake of speed is valueless and self-indulgent.

Speed and cost reduction are transformed into commodities by modeling media. Much as spreadsheet software commoditized financial modeling, digital design media commoditize cycle-time compression. Faster development cycles are inevitable. Thus developing ever faster at lower cost isn't the issue; the issue is getting greater value from the time and money saved. What happens to competitive advantage as cycle-time differences between rival firms narrow and their offerings hit the market simultaneously? What happens when development cycles are shorter than customers are able to absorb?

The challenge is to treat speed, cycle-time compression, and their concomitant savings less as ultimate ends and more as a creative means. Speed-to-market isn't merely a source of competitive advantage; it is a resource enabling innovation and differentiation. When organizations rethink the role of speed in shaping their value propositions, they invariably discover that new modeling media demand new behaviors. Traditional distinctions between tactical and strategic trade-offs blur. Organizations find they must constantly ask themselves hard questions and then decide whether or not to reorganize around their answers.

Prototyping cycles—we could call them virtual rehearsals or seamless simulations—are ultimately a form of innovation *capital*. How a firm invests that capital reflects how it reconciles its own values with what the marketplace values. More cycles mean more capital, which means more choices. The more choices you have, the more your values matter. The more your values matter, the surer you must be of why you hold them so dear—or why they must be discarded.

> Developing faster and cheaper isn't the issue; it's getting greater value from the time and money saved.

A simple scenario illustrates the tensions and trade-offs. A new rapid-prototyping-and-seamless-simulation infrastructure enables a company to double the number of development cycles its product teams can run. Under the old system, a new-product team could perform ten cycles during its ten-month development window. Now, for the same cost, the team can run twenty full iterations. Think of the extra cycles as currency: each additional cycle can "purchase" a product improvement, a cost reduction, or a speedup. All cycles are equally valuable. Unspent cycles are monies saved.

How should a team spend or invest those ten extra cycles? What expenditure will give the best returns? Should the team

- spend all its cycles on speed, to come to market in half the time?

- spend its cycles on improvements, and come to market after ten months with a product that's 50 percent "better"?

- spend its cycles on cost reduction, to be able to cut prices 30 percent?

- spend all ten cycles on the ideal blend of speed, price, and quality? If so, what is that optimal blend?

- bet a couple of cycles on an intriguing but risky enhancement?

- use a few cycles to test an alternative design approach?

- save two or three cycles to keep development costs down?

- use all ten cycles to develop an entirely new product concept?

There is no inherently right answer. Even worse, these questions are too simplistic. They lack the menace and nuances of organizational conflict that managers confront when hard choices have to be made. Should the development team use the extra cycles to focus on particular product features or specific cost reductions? How will it allocate the extra cycles? Should design, manufacturing, and marketing each get three cycles, with the tenth held in reserve for emergencies? Or does the product manager "own" the cycles budget? At what point should key customers and suppliers be brought in to help spend extra cycles? from the very beginning? toward the end? which customers and suppliers? Or should doubling the number of development cycles have no impact on how the firm manages its design relationships with suppliers and customers?

Spending new cycles could be a greater challenge than investing new money.

Productively spending new cycles could prove a greater management challenge than successfully investing new money. Development cycles aren't quite as fungible as cash. The ability to prototype more options in less time ultimately becomes an entirely new organizing principle for managing value creation. Or, alternatively, there is the danger of falling prey to a "Parkinson's Law of Prototyping," whereby endless iterations soak up available time while offering little but diminishing returns.

Is the 50th iteration of a prototype or simulation as valuable as the 35th? the 60th? the 110th? Or are the insights and information gleaned merely marginal? Do the design discussions fundamentally shift? Or do they simply become more refined? Do design assumptions harden? Or do they become less constraining as the cost of testing them drops? These questions are not hypothetical.

They are at the white-hot center of management decision making. Are we creating value? Or are we just futzing around?

Waste as thrift

George Gilder, a rhapsodic commentator on technology, persuasively argues that there come moments when innovation vaporizes economic assumptions. "From time to time," he writes, "the structure of nations and economies goes through a technological wringer. A new invention radically reduces the price of a key factor of production and precipitates an industrial revolution. Before long, every competitive business in the economy must wring out the residue of the old costs and customs from all its products and practices."

Gilder's favorite example is the integrated circuit—the microchip that reduced the price of electronic circuitry by a factor of over 1 million. That technology-driven economic transformation completely changed the silicon design paradigm. Electronics designers were forced to put all their traditional design methodologies through the technological wringer. "Electronic designers now treat transistors as virtually free," Gilder writes. "Indeed, on memory chips, they cost some 400 millionths of a cent. To waste time or battery power or radio frequencies may be culpable acts, but to waste transistors is the essence of thrift. Today you use millions of them slightly to enhance your TV picture or to play a game of solitaire or to fax Doonesbury to Grandma. If you do not use transistors in your cars, your offices, your telephone systems, your design centers, your factories, your farm gear, or your missiles, you go out of business. If you don't waste transistors, your cost structure will cripple you. Your product will be either too expensive, too slow, too late, or too low in quality."

To waste transistors is the essence of thrift.

To waste transistors is the essence of thrift. This assertion raises a related question: If "free" digital models are to product, process, and service design what "free" transistors are to electronic design, could wasting cycles be the essence of thrift? Consider this play on Gilder's words: "If you don't waste simulations and prototypes, your cost structure will cripple you. Your product will be either too expensive, too slow, too late, or too low in quality." Ingeniously "wasting" prototypes is therefore essential to risk management. Throwing simulations at design problems becomes vital both to detecting errors and to discovering opportunities. Failure then to wring new productivity from such riches isn't just an embarrassment; it represents a lost opportunity of epic proportions.

The Cultural Economics of Waste Management

Economist and Nobel laureate Ronald Coase has pointed out how digital technologies can dramatically reduce organizations' internal coordination and external transaction costs. Coase's insights merit special attention; he won his Nobel for his precomputer insights into the nature of the business firm.

Coase asked the simplest of questions: Why do firms exist at all? His answer, which gave rise to the discipline of transaction-cost economics, examined how firms decide which aspects of their operations they handle internally (incurring coordination costs) and which they decide to buy (incurring transaction costs). Make-versus-buy decisions are the foundation of the theory, which casts the firm and the marketplace as alternative—even rival—venues for the very same activities.

Managerial champions of "core competences" and "outsourcing" are Coase's disciples. They seek to outsource activities that the market can provide more cost-effectively than the firm does, reducing their internal coordination costs to make their core competences even more valuable. At the largest companies, top managers wage a constant battle to assure that coordination costs don't outweigh economies of scale.

Blurring make-versus-buy decisions

Discontinuities in coordination costs or transaction costs fundamentally change a firm's make-or-buy decisions. That's what the digital revolution has been all about: using ever-cheaper processors and networks to help organizations slash their costs. Digital media have clearly cut internal coordination costs for prototyping. Coase emphasizes that digital technologies have also reduced the external costs of transacting with suppliers. That's why product-development cycles get faster and cheaper.

But cutting old costs can create new complexity. What happens to the nature of the firm when managers can't distinguish between coordination costs and transaction costs? What happens when digital modeling media blur the make-versus-buy distinction? Who benefits? Who pays? The rise of digital modeling media is creating markets in which it's increasingly unclear who really runs the development-cycles portfolio.

What happens to the nature of the firm when managers can't tell coordination costs from transaction costs?

Consider Chrysler, Ford, and Boeing. Each insisted that its key suppliers adopt its CAD/CAE digital infrastructure. All insisted that their suppliers become even more responsive, cost-conscious, and innovative. Chrysler wasn't satisfied in 1995 when it standardized its six thousand engineers on CATIA; the firm wanted its suppliers on CATIA too. Chrysler back then purchased about 70 percent of the content of every car from suppliers and had multibillion-dollar transaction costs. "We told all our suppliers: 'Not only do we want you in full partnership, we want you to be electronically linked,'" says John C. Miller, general manager of Chrysler's large-car platform engineering. As an incentive, Chrysler told its suppliers that they could keep half of any resulting cost savings.

Was linking Chrysler's suppliers into its digital design infrastructure a make-or-buy decision? a make-and-buy decision? or something else? When Chrysler rapid-prototypes with its most important suppliers, is the firm managing coordination costs or paying transaction costs? Or have Chrysler's suppliers been coopted into subsidizing one of their biggest customers? Are Chrysler's networked suppliers now insiders, outsiders, or something else? What would Coase say?

Coase would predict that dramatic declines in either coordination costs or transactions costs should create new markets in relationships, both inside and outside the firm.

Declining coordination costs enhance opportunities for collaborative interdepartmental and cross-functional initiatives. Chrysler's platform teams and Boeing's design/build teams both embody this principle.

> **Declining coordination costs promote cross-functional initiatives.**

Decrease the coordination costs, and the firm lowers its barriers of entry for greater internal interaction. You can't manage cross-functional teams without cross-functional prototypes. Prototypes enable collaboration. Declining transaction costs promote multiple dealings with multiple suppliers—or the conversion of transaction costs into coordination costs by linking suppliers into internal networks. Look at the real world: this is precisely what has happened in computer-aided design-and-manufacturing networks. New relationships are being created around the simultaneous plunge of transaction and coordination costs.

Iterate, innovate, or evaporate

As firms integrate external suppliers into their internal modeling marketplaces, the economic issue shifts from the classic make-versus-buy decision toward another kind of imperative: "Iterate, innovate, or evaporate." The economic relationship becomes contingent on the supplier's perceived ability to create value collaboratively, not just to sell it. Are the modeling cultures of the networked firms compatible? Before the surge in networked CAD/CAE, that was a silly question. In the Internet era, the only question that matters more is whether the modelers are competent.

With reduced transaction and coordination costs, networked firms now negotiate relationships characterized by:

- iteration, which calls for coordination with only marginal changes, or

- innovation, which entails improvements and/or cost reductions that create new value, or

- evaporation, which involves either subsuming the supplier or its components into a larger system or purging them from the iterative process as free riders.

As lower costs restructure the firm's transactions and its coordination economics, managers increasingly have to address two key strategic questions:

- Who else should we be innovating with?
- How should we be innovating with them?

Theoretically, negligible transaction costs could enable a firm to network with dozens of suppliers of a particular component. If the component is a pure, unadulterated commodity, electronic networks can create auction environments in which vendors bid for the right to sell to the firm; General Electric's purchasing intranet is a first-rate example. The firm's procurement arm becomes a bit like the NASDAQ trading floor.

Partnering and customization

If a firm counts on its suppliers to do more than merely build a component to specifications—if it wants suppliers to suggest enhancements, innovations, and price reductions—it has to be very selective. Although, technologically, the **A networked firm has a strong incentive to consolidate suppliers.** firm could link a dozen suppliers to its network to help codesign a particular widget, it probably lacks the time and talent to juggle a dozen development proposals for a single component. Multiply that by the thousands of components and scores of subsystems that go into a typical automobile, and the conclusion is obvious: the networked firm has a strong incentive to consolidate the number of its collaborative suppliers when it wants to increase its rate of innovation and differentiation.

In fact, global manufacturers that have committed to cutting-edge digital infrastructures—the Chryslers, Boeings, Hewlett-Packards, and IBMs—have all consolidated the number of suppliers with whom they codevelop products and services. The dramatic drop in transactions and coordination costs economically encourages companies to look for new partnerships to cocreate new value.

Vendors once dominated their markets by offering the right products at the right price. In the era of digital modeling, successful vendors have to be organizations with whom their customers want to waste design and development cycles together. Dell Computer's popular and profitable Web-based "premiere pages" illustrate the point. Dell offers its corporate customers the opportunity to "test-model" various configurations of the personal computers they plan to buy. Publicly, this interface is positioned as a case study in the power of "mass customization." This is not quite accurate: the decision to turn the customized configuration into a purchase is reached only after numerous iterations exploring various price and feature configurations. The value isn't just in the customization; it's in the medium and the process of customization. Dell's biggest customers are encouraged to "waste" customization cycles so that they can be confident that the configurations they ultimately choose are the right ones.

> **Sophisticated clients don't just want customized products and services; they want the power to customize.**

Leaders in this "product-configuration" software space, such as Calico Commerce and Trilogy, are constantly looking for ways to boost the modeling-and-simulation components of their offerings. Not just because that's what their customers want; it's what their customers' customers demand. Sophisticated clients don't just want customized products and services; they want the *power to customize*. Customization isn't choosing from a menu; it's the ability to test any recipe. So it's no longer enough even to design and deliver innovations to specifications;

suppliers must now be able to codesign and coproduce innovations to specs that change with every new cycle. Discerning the differences between transaction costs and coordination costs in this environment is a task best left to Nobel laureate economists.

Symbiotic design relationships

Should the supplier take the lead in suggesting innovation, or does the customer call the shots? Does the customer consider diminishing returns to have set in somewhere around the fiftieth design iteration while the supplier doesn't get antsy until the eightieth? Does a firm that does ten iterations a week have a client that wants to do twenty? Different firms have different development rhythms. Some customers respond well to innovative suggestions; others do not. Does the supplier spend lots of cycles creating multiple options with the client? Or do their collaborations revolve around designing the "best" option? Are there constant customer/supplier struggles between prototyping to cut costs and prototyping to add value? How is that conflict managed?

Companies like Sharp and Chrysler are confident that their modeling infrastructures will bring products to market faster and cheaper. The more their confidence grows, the more likely it is that management's attention will shift to managing the process: we know we're going to be 35 percent faster and 25 percent cheaper, so what else are we going to be? The focus shifts to the kind of product that will be made 35 percent faster and 25 percent more cheaply. How will cycles be managed within the context of these known outputs?

The "hows" all revolve around the new economics of wasting cycles. The economist and Nobel laureate Herbert Simon, author of the classic design text *The Sciences of the Artificial*, observed decades ago that digital media radically reduce the costs of searching for solutions to a problem. The ability to conduct low-cost searches for solutions also transforms design relationships, reinforcing the effects of plummeting transaction and coordination costs.

At suppliers like ITT Automotive, the ability to create virtual prototypes profoundly alters the definition of a component. Instead of optimizing the design of individual parts, suppliers and customers now search for opportunities to cocreate subassemblies and subsystems that totally eliminate the need for particular parts. The ability to play with virtual prototypes makes it far easier to explore design-for-assembly/design-for-manufacture opportunities. Simulating the subsystem can become more valuable than prototyping the part. Suppliers and customers can then spend their cycles figuring out what the appropriate unit for prototyping should be. Is it the part? the subassembly? the subsystem? the interface between the component and the system? Japan's Canon has collaborated with its suppliers to slice over one-third of the components out of its printers in under three years. Pirelli, the giant Italian tire company, actively collaborates with European auto manufacturers to integrate its tires into suspension systems in ways that cut parts and weight. Pirelli constantly pushes for greater influence over the design of suspension systems on the strength of its superior knowledge of what happens when the rubber meets the road. Ideally, these design relationships prove symbiotic: the enterprise buys fewer components, and the vendor moves up the food chain by providing a value-added subsystem customized for the ultimate product. Ultimately, customers and suppliers are investing cycles in discovering the kind of business relationship they should have.

Transparency and Bozo Filters

Design relationships aren't always mutually beneficial. The same digital modeling media that facilitate collaboration can also be used to "evaporate" a component and the supplier who makes it.

Digital modeling media like CATIA inevitably make the prototyping process more transparent. CAD/CAE software makes it possible to examine the design assumptions of a prototype, and networked digital media make suppliers' design assumptions more accessible to their customers, and vice versa. Greater openness always forces the parties in a relationship to reexamine how they should interact with each other. What should be shared? What must be concealed? How safe is it to collaborate? What are the rules of engagement?

Technology-driven transparency allows for aggressive forms of relationships management. For example, many electronic bulletin-board and e-mail systems permit what the "digerati" call "bozo filters." Bozo filters screen out contributions from people whose insights and opinions users consider worthless. In many Internet newsgroups and user groups, bozo filters have become a popular way to help manage information overload.

Technology-driven transparency allows for aggressive forms of relationships management.

In a corporate context, the use of bozo filters could be devastating. What happens when manufacturing managers decide to bozo-filter the modifications of certain design engineers? What happens if European managers decide that the Asian regional managers' contributions are more trouble than they're worth to evaluate? What happens when the vice president of marketing is rumored to be bozo-filtering the design critiques of the chief engineer? Or when the California design group bozo-filters aesthetic enhancements submitted on the CAD network by the Milan design staff? What should a supplier do if it discovers that a key customer surreptitiously filters its designs?

This simple example illustrates the difficulty of managing relationships in an era of technology-driven transparency. An excellent case could be made for either encouraging the use of bozo filters or prohibiting them. The choice depends on what kind of relationships an organization wants to have. Just as firms have different design sensibilities and appetites for risk, they also assign different values to technology-defined transparency in relationships.

This brings us full circle: a firm's design partners shape how it codesigns; how a firm codesigns may determine whom it chooses as design partners. Changing the economics of modeling inevitably changes the politics and culture of modeling. As development cycles become more of a commodity, partnerships and processes become less so. And the lower the costs of search, transactions, and coordination, the more valuable design relationships seem to become.

As development cycles become more of a commodity, partnerships and processes become less so.

The easier it becomes for firms to waste their development cycles, the more important it becomes to determine with whom and how best to waste them.

Model Risk

Up to now we have been making a dangerous assumption: that the models work. The history of software development—and that of digital prototypes—is a history of bugs and flaws. Nor are the models themselves risk free. Just as an error in a spreadsheet simulation can have catastrophic financial results, errors in CAD/CAE design simulations can result in crippled products. An inadvertent sign-flipping from a plus to a minus can prove catastrophic.

We have been making a dangerous assumption: that models work.

Several disturbing experiments indicate that naive students and even sophisticated engineers place too much trust in calculations performed by computer. That's why many organizations run their calculations and simulations on several brands of software: to assure that significant errors built into the models will surface.

Even so, *model risk* is often woefully underappreciated and undermanaged. Since the era of ancient Greece, engineers have known that models often don't scale well: structures simply collapsed when built full-size. In the digital domain, managerial modelers have discovered that risk doesn't always scale well; reality's probabilities frequently prove more vicious than the model's probabilities.

"Every model has an Achilles' heel," asserts Tanya Beder of Capital Markets Risk Advisors, "and the question is, How do you shore it up? With a cast, a splint or a cane? . . . When life is boring, there's not a lot of difference in models. But when markets are volatile, they can be wildly different."

The models Beder runs are used by hedge funds and other financial firms to manage their portfolios. These sophisticated instruments often consist of multibillion-dollar bets based on financial models that predict mathematically "correct" relationships between various options, derivatives, swaps, and synthetics. When these models work, financial institutions can reap hundreds of millions of dollars in a single trade. When they fail, as they did in the case of Long Term Capital Management; NatWest; D. E. Shaw; Kidder, Peabody; and a host of other quant-driven traders, billions of dollars can be lost. Bankruptcy joins the list of options.

"These firms have to ask themselves, How did we *really* make our money? What kind of risk-dollars did we spend? . . . Who spent more risk? Why?" says Beder, whose firm works closely with financial institutions and their boards to assess their models and the risks they take. According to Beder, too few financial institutions are aware of their level of risk exposure. Complicating the situation, she adds, is that the models used to manage risk are often risky themselves.

Unlike CATIA, which must respect the laws of physics, the models built by financial engineers must balance the science of finance with the sociologies of markets. Measuring risk in a way that can be meaningfully managed becomes extraordinarily difficult. In this respect, risk management in financial markets represents a microcosm of the key challenge facing all organizations that seek to model their own "risk appetites" in their respective markets.

Mark-to-market mishaps

Beder and other sophisticated modelers warn in particular against letting models become a substitute for the marketplace. The result is a "mark-to-model" pathology: the firm believes that, as long as the model is behaving according to its parameters, its risk is being managed. Dealers think they are marking their positions to the market when all they are really doing is marking them to the assumptions built into the model. "At least six disparate models are used to price many common derivatives," Beder has observed. "Given the same raw data input, a wide variance of results exists when mark-to-market is calculated." According to one survey, valuations varied as much as 6 percent on relatively plain-vanilla transactions. For more complex derivatives, the valuation spreads can jump to double digits. Many banks have set aside reserves for model risk—for when their models are out of whack with the market.

Sophisticated modelers warn against letting models substitute for the marketplace.

In 1996, the Bank of Tokyo-Mitsubishi in New York discovered that a model used by one of its options traders was at odds with the market to the tune of $50 million at a mid market price. An additional $33 million was lost when senior management back in Japan decided to close out the position. "There was no violation of the rules," Tokyo-Mitsubishi International chief executive Akira Watanabe told *Euromoney*. "Market volatilities were changing, and they reviewed the options models they were using to be more conservative."

A mark-to-market mishap appears also to have been responsible for NatWest's $150-million options-trading losses in 1997. The bank preferred to blame a rogue trader, but the market consensus is that the bank traded mispriced options for a sustained period. Practitioners insist that almost every case of options misvaluation—there have been billions of dollars' worth of them—boils down to the same fundamental mismanagement: allowing traders, directly or indirectly, to determine their own valuations, or, in other words, to mark to models of their own making.

This litany of modeling and simulation disasters applies frighteningly well to other industries. Do engineers "mark to model" by failing to test key assumptions with suppliers and customers? How do managers test their prototyping assumptions in the real world? What methods and methodologies are used for verification and validity? Who audits the models and simulations and prototypes? When?

Stress-testing models

"Derivatives doctors" like Tanya Beder push for independent risk oversight to "stress-test" models for major flaws: in the case of exotic instruments, wrong assumptions about the components of complex derivatives, illiquidity and poor correlations between products thought to be similar, and legal uncertainties. Typically, financial organizations stress-test their models by comparing them to historical data, trotting them through worst-case-scenario analyses, and subjecting them to Monte Carlo–method simulations.

One problem, notes Beder, is that historical performance data are notoriously unreliable as parameters for future financial-markets behavior. "Much of the historical data these models have been built on now appears irrelevant," she says. "The past has been a horribly misleading guide to the future." Scenarios are by nature constrained to what their conjurers choose to imagine. Surprises are, by their nature, unanticipated.

"The past has been a horribly misleading guide to the future."

As for Monte Carlo methods, one risk-management software developer observes, "If you design your Monte Carlo simulation to show you the worst-case scenario, you could end up managing by fear and missing trading opportunities in constant expectation of the next market crash." No matter which method of stress-testing model assumptions an organization chooses, in other words, there are risks to be managed.

Again, the issues in the model-driven marketplaces of financial management mirror the modeling challenges confronting innovators. Who are the "independent risk overseers" for the digital prototypes the firm uses to manage innovation? To what stress tests are prototypes subjected? What role do customers and clients play? Do they help generate scenarios? Or is the firm so comfortable with its prototyping processes that marking-to-model is good enough? In essence, what is the risk appetite of the firm when using models to manage the risks and opportunities of innovation?

One purpose of prototypes should be to create conversations and interactions that ultimately shrink risk rather than exacerbate it. Confusing model risk with market risk is perhaps the most dangerous risk a firm can make. But as organizations exploit the opportunity to do more modeling for less, the level of model risk inevitably climbs.

5

PREPARING FOR SURPRISE

You must be prepared for a surprise,
and a very great surprise.
Niels Bohr

During the deregulation of U.S. airlines in the 1980s, the government and the airlines swore that one core regulatory value remained sacrosanct: safety. But fervent promises that deregulated flight posed no new risks had to be balanced against the new marketplace reality that planes and their pilots were working longer hours. Given the increasing length and frequency of transoceanic flights, and the often-erratic variability of pilot schedules, pilot fatigue was seen as a potentially serious problem. Could tired pilots make fatal mistakes? What role did fatigue play in pilot effectiveness? How much fatigue is too much? How much is "safe"? A nervous Congress wanted answers. Human-factors researchers at the National Aeronautics and Space Administration (NASA) were told to investigate the significance of pilot fatigue in aircraft operations.

These questions could hardly be safely tested in the real world. Simulation, the researchers decided, was the appropriate medium for assessing pilot performance. The researchers settled on a Boeing 737 simulator to examine how well experienced two-person cockpit crews would respond to a particularly wicked but realistic flight scenario.

The researchers tested two groups of test crews: those who flew the scenario after a minimum of two days off, as if it were the first leg of a three-day trip (preduty) and those who flew the scenario as the last segment of a three-day trip (postduty). The scenario was characterized by poor weather that forced a missed approach to a landing. The missed approach was further complicated by a hydraulic-system failure that created a high-speed, high-workload situation. The two pilots had to select an alternate landing site and manually extend the plane's gears and flaps while flying an approach at higher-than-normal speed.

Fatigued crews perform better than rested crews.

As expected, the postduty crews had had less presimulation sleep and reported significantly more fatigue. But, to the researchers' astonishment, "fatigued crews were rated as performing significantly better and made fewer serious operational errors than the rested, preduty crews."

As NASA's researchers commented, "In hindsight, the finding shouldn't have been a surprise at all. By the very nature of the scheduling, most crews in the postduty condition had just completed three days of operation as a team. By contrast, those in the preduty condition normally did not have the benefit of recent experience with their other crew members."

When the researchers reanalyzed their data, fatigue was found to be a far less statistically significant safety factor than whether the crews had recently flown together. The simulation findings indicate that crew schedules resulting in frequent mixing of pilot teams can have significant operational implications. The NASA researchers noted that no fewer than three of the worst 1980s-era accidents—a stall under icy conditions, an aborted takeoff that landed the plane in the water, and a runway collision in dense fog—all involved crews paired for the first time.

Did NASA's findings imply that pilot fatigue was irrelevant to flight safety? Of course not. But the simulations uncovered compelling and unexpected management issues. Rigorously managing

pilot pairings may prove far more important for flight safety than
rigorously managing pilot hours. U.S. air-
lines took note: airlines now monitor and
manage pilot pairings. That this life-or-
death management issue was discovered
by accident speaks volumes about the
challenge of creatively managing prototypes and simulations.

> **This life-or-death management issue was discovered by accident.**

Harnessing Surprise

The real value of a model or simulation may stem less from its
ability to test a hypothesis than from its power to generate useful
surprise. Louis Pasteur once remarked that "chance favors the prepared mind." It holds equally true that chance favors the prepared
prototype: models and simulations can and should be media to create and capture surprise and serendipity. Yet surprises are not
always welcome. More often than not, prototypes and simulations
are designed to eliminate surprises, not create them. Models are
usually built to test ideas rather than generate new ones and risk
management is emphatically not the same thing as opportunity creation. Prototypes and simulations are rarely designed to contradict
the underlying assumptions they embody.

Interpreting the unexpected

When such contradictions occur, do organizations ignore the
model's results or accept them? Do they change the model's
assumptions? How is surprise managed or mismanaged? Clark Abt
of Abt Associates, a pioneer in applying simulation games to public
policy, recalls running a simulation for the Agency for International
Development (AID) involving sustainable economic development
in a developing country. "The simulation was biased in favor of saving the forests, while still allowing for a growing population and

increasing the standard of living," Abt recalls. The overt goal was, in his words, "to learn how to save the environment in a politically responsible way while having healthy economic development." But practically every run of every simulation led to the relatively rapid destruction of the ecologically cherished but commercially irresistible forests. "By the end of the day, the forests were all gone," Abt remembers. "The AID types were really pissed off."

So what did AID do in the ugly face of this consistent and politically incorrect outcome? The agency shut down the exercise. The hoped-for outcome that this country could manage to avoid painful trade-offs between economics and the environment simply wasn't true.

"You know you have something when the model has a life of its own."

Yet this story illustrates the intellectual and emotional power of models and simulations. "You know you have something," Abt asserts, "when the model has a life of its own."

This ability to startle and surprise—both pleasantly and not—is what animates a model. A "personality" emerges. The model becomes, in effect, a character with its own idiosyncrasies and insights. It elicits human interaction and creates interactions between others. People have to choose which aspects of the model's character they will try to understand and which aspects they will deliberately ignore. They have to decide which surprises are welcome and which represent a blasphemous breach of faith. The NASA and AID simulations generated insights that went well beyond the original premises they were designed to test, forcing the "experts" to rethink their most fundamental assumptions.

Designing to convince

Designing evocative models presents a straightforward trade-off between their ability to surprise and their capacity to be understood. Abt glibly compares the design of models to women's skirts: "They should be long enough to cover the subject but short enough to be interesting." To put it another way, the most potent models are complex enough to yield counterintuitive results yet simple enough to let people grasp how those results were obtained. Models whose results merely confirm expectations are redundant; models whose results are surprising but inexplicable provoke without context. The challenge is to devise transparent models that also make people shake their heads and say "Wow!"

> The challenge is to devise transparent models that also make people say "Wow!"

The most sophisticated model designs rarely prove capable of provoking surprise, while the simplest models often generate reams of counterintuitive insights. High-fidelity models and simulations can evoke the "suspension of disbelief" that makes the best movies and fiction seem real. On the other hand, even a few crude equations or spreadsheet plottings on a personal computer can produce results that startle sensibilities. Both kinds of models may belong in an organization's prototyping portfolio. But, as the computational costs of model-building continue to plummet, which design approach represents the better value? As organizations struggle to manage complexity, which approach is likelier to evoke useful behaviors? For example, the high-fidelity realism of cockpit flight simulators clearly offers benefits that a low-fidelity videogame-like flight simulator would lack. But is it really necessary to faithfully recreate a country's economy to explore the implications of sustainable-development policies? Precisely what questions are we modeling for?

In a 1994 article in *Science*, three scientists argue that even the most sophisticated mathematical models are inadequate to explain the natural physical phenomena they seek to simulate. Scientific debate about issues such as global warming, they argue, has become far too dependent on models whose explanatory powers are questionable at best and dangerously misleading at worst. "Verification and validation of numerical models of natural systems is impossible," they assert. A far better case could be made for the impossibility of verifying and validating numerical models of business systems: human foibles are far harder to model faithfully than physical ones. But the article offers an important observation that maps directly onto the businesscape. The authors quote philosopher Nancy Cartwright's characterization of numerical models as "a work of fiction," and then comment:

> *While not necessarily accepting her viewpoint, we might ponder this aspect of it: A model, like a novel, may resonate with nature but it is not a "real" thing. Like a novel, a model may be convincing—it may ring true if it is consistent with our experience of the natural world. But just as we may wonder how much the characters of a novel are drawn from real life and how much is artifice, we might ask the same of a model: How much is based on observation and measurement of accessible phenomena, how much is based on informed judgment and how much is convenience? . . . [We] must admit that a model may confirm our biases and support incorrect intuitions. Therefore, models are most useful when they are used to challenge existing formulations rather than to validate or verify them.*

This notion that models and simulations serve us best when they challenge our assumptions has profound implications for innovation management. The ability to respect the counterintuitive and embrace the unexpected is essential both to managing risk and to creating opportunity. In practice, creative counterintuition drives innovation.

Models serve us best when they challenge our assumptions.

Yield Management: Beyond Counterintuition

Marriott International, like so many other service companies looking to strike a more profitable relationship between pricing and inventory, turned to yield-management techniques to rethink how to make the most money per guest per room. Marriott's key operating metric was "revenue per available room," and the company managed itself accordingly.

Yield management—also known as revenue management—is a mathematically intensive approach to forecasting consumer-demand patterns and optimizing the mix of product availability and price accordingly. Airlines, hotels, rental-car companies, and utilities have all turned to yield management to redefine their rules of revenue generation. When a plane lifts off with empty seats, or a hotel room goes unfilled, or a rental car goes unrented over a weekend, their potential revenues are forever lost. Yield-management models offer clever ways to match pricing structures to perishable products. Companies that use yield-management techniques effectively have reported 3 to 7 percent topline revenue increases. Growth of this magnitude frequently translates into profit increases of 50 to 100 percent. Yield-management modeling has proven to be a remarkably cost-effective simulation platform for simultaneously boosting revenues and margins.

Nevertheless, phasing in a yield-management system often proves to be an exercise in managerial resistance. At Marriott, the statistical models of anticipated customer demand directly challenged managers' claims to expertise. Marriott's CEO J. Willard Marriott offers a telling anecdote:

> [Our] Demand Forecasting System (DFS) concept was met with a lot of skepticism, especially from "old timers" who felt that guest behavior could not be predicted, or if it could be, that they could do it better than a machine. But testing proved them wrong.

> In an early test phase, conventional wisdom held that no discount rates whatsoever should be given at the Munich Marriott during Oktoberfest because of the tremendous demand. Yet the DFS recommended that the hotel offer some rooms at a discount, but only for those guests who would stay for an extended period either before or after the peak celebration days.

> Although counter to what the [hotel's] general manager felt was common sense, he applied the DFS recommendation and was

pleasantly surprised with the results at the end of Oktoberfest.
Although the average daily rate was down 11.7 percent for the period,
occupancy was up more than 20 percent and overall revenues were
up 12.3 percent.

Good for Marriott. But the point isn't that yield management
worked; it's that the CEO claims that it worked at the expense of
Marriott's "conventional wisdom" and "common sense." The innovative power of the model arose from its
Yield management   ability to challenge assumptions rather
worked at the expense   than reinforce them. As Will Rogers put it,
of conventional   "It's not what you don't know that hurts
wisdom.   you, it's what you do know that ain't so." As
valuable as executive experience and intu-
itions may be, Marriott now insists they be tested in the simulated
marketplace of its yield-management algorithms.

At firms like American Airlines and Hertz, yield-management
simulations have rapidly evolved from "counterintuition engines"
into a form of CAD software for retooling the business model.
Much as digital simulation transformed how Boeing built its 777,
it also transformed how American Airlines slots, schedules, and
prices its flights. No longer limited to quantitatively modeling
inventory management, these simulations play a growing role in
capacity expansion, resource allocation, and promotional currencies like frequent-flier miles. The algorithms let managers play with
prices and inventory optimized around a host of different variables.
Not unlike a spreadsheet model, a yield-management simulation
encourages what-ifs.

Marriott could explore how its pricing might change if it shifted
its key metric from "revenue per available room" to "profit per
property per day." How might that shift affect a manager's desire
to attract more corporate meetings on-site? American Airlines
could explore whether tickets blending frequent-flier miles with
real money might boost revenues and margins from the airline's
most frequent fliers. Will the answers to such hypotheticals be

"common-sensical"? Or will they be surprises that raise new questions and inspire new marketing experiments?

In effect, yield-management simulations—tightly coupled with data-rich revenue models—create a virtual marketplace for profitable surprise. The beauty of these simulated marketplaces is that they are congruent with the real-world marketplace. The critical question is how well the simulated surprises map to reality. And what role can these models and simulations play in managing the more perverse counterintuitive results and surprises found all too frequently in real-world design? The counterintuitive challenge provides a reality check.

Yield-management simulations create a virtual marketplace for profitable surprise.

Emulating Reality

Traffic jams represent idleness, waste, and pollution on a worldwide scale. From Boston to Bangkok, tens of billions of dollars are invested in new highways to smooth the flow and accelerate the pace of automobile travel.

The statistical models that have been used to predict traffic patterns typically treated traffic as if it were a problem in fluid dynamics. They did a far better job of simulating ideal traffic flows than anticipating turbulence. But new mathematical techniques are transforming how traffic gets simulated. Billions of dollars in productivity and billions of hours in traffic jams depend on how these emerging simulations are interpreted and applied.

"Transportation is in a very cool spot between a social system and a physical system," Christopher L. Barrett, who studies traffic at Los Alamos National Laboratory, told *Scientific American*. Because human behavior behind the wheel sometimes defies physical laws as well as traffic laws, even the most logical solutions to traffic problems can have counterintuitive consequences. For example, traffic engineers and frustrated commuters alike have long believed that the surest way to dilute traffic jams and boost traveling speeds is to build more capacity: expand the number of lanes, or build another roadway or two.

Even the most logical solutions to traffic problems can have counterintuitive consequences.

Alas, reality tells a different story. To the surprise of planners and to the dismay of commuters, new roads fail to ease peak-time traffic jams. In fact, they often make them far worse. The plans and statistical models that have justified new taxes and bond issues to build new roads have proven utterly false. Expensive roadways built to make people's lives easier frequently have the opposite effect. These obnoxiously perverse results—which have been observed worldwide—have a name: Braess's Paradox, named for the German operations researcher who first articulated the phenomenon in 1968. Braess discovered that raising a network's traffic capacity often slows average travel speed. Adding a lane or a new route can create bottlenecks when too many drivers surge onto the new shortcut. These rush-hour bottlenecks often end up creating slowdowns and gridlock elsewhere in the traffic network as well. As Steen Rasmussen of Los Alamos National Laboratory points out, "When you design roads, you want to maximize throughput [traffic flow]. But, it turns out, at the point of most throughput predictability drops. This means as you go toward capacity, reliability of the traffic system breaks down." The goal, then, is to design systems that function just under capacity.

This finding was utterly counterintuitive and cut sharply against the generational grain of opera-

Simulations can link physical-system behavior to social-system behavior.

tions research and traffic-engineering tradition. Braess's Paradox forced highway planners around the globe to rethink roadway design, construction, and management. A new generation of simulations has gradually enabled traffic planners to model traffic behavior more realistically. Instead of treating traffic as a physical flow of automobiles across a network, they can emulate the behavior of thousands of individual cars. Physical-system behavior can be linked to social-systems behavior. The unit of simulation has shifted from traffic to cars. Braess's Paradox can now be effectively simulated out of existence before the first shovelful of earth is turned over.

Braess's Paradox represents a case study in creative dialogue between learning from reality and learning from models. Organizations can't manage Braess's Paradox and its counterparts without both a willingness and an ability to treat counterintuitive results as a challenge rather than an aberration. Models that can emulate and explain counterintuitive phenomena expand the boundaries of thought.

Customized Surprise

Confounding the experts

The notion of counterintuitive surprise raises a related question: Who is being surprised? In the preceding examples, "the experts" were surprised. The behavior of the models forced them to change their own mental models and rewrite the rules of the game. Their "expertise" made them hypervulnerable to surprise but well situated to turn surprise to their advantage. That's why Alexander Fleming recognized the importance of a mold on an agar plate and discovered penicillin. And how Charles Goodyear recognized the serendipitously discovered properties of vulcanized rubber. Even software errors can yield design breakthroughs. "Bugs are an unintended source of inspiration," asserts Will Wright, the creator of globally popular simulation games like Sim City. "Many times, I've seen a bug in a game and thought, 'That's cool; I wouldn't have thought of that in a million years.'" Chance does favor the prepared mind. The question is, What is that mind prepared to perceive? And in what form does chance, or surprise, make its appearance? In science, technology, and the arts, it often appears in prototypes and experiments undertaken as media for discovery.

Sometimes, who isn't surprised proves as interesting as who is. Perhaps the pilots could have told the NASA researchers that too many partners posed a greater threat to flight safety than too many hours in flight. Perhaps any airline with a healthy yield-management operation could have told Marriott's managers that their gut instincts were wrong. Perhaps any commuter could have told highway-traffic engineers that their expensive new lanes and roadways were leading to bigger traffic jams. That's another reason why models and simulations are instruments for introspection: the organization can learn from the conversations that otherwise wouldn't take place which surprises and epiphanies are unsurprising.

This element of sophisticated surprise and counterintuition is what makes innovation so tricky. Expertise does not confer wisdom. A client, a customer, a marketplace may not want surprise or novelty. They may prefer predictability. Or they may simply want their problem solved as swiftly, painlessly, and cheaply as possible. What value does modeling and simulation for surprise have for them?

Demand articulation: Knowing it when you see it

Japanese management scientist Fumio Kodama coined the term *demand articulation* for the process whereby consumers of innovation discover—rather than know—what new products and services they need. In fast-moving markets, customers and clients are often unable to articulate what they want and need. They know it when they see it.

> **Consumers of innovation discover— rather than know—what new products and services they need.**

They also know it when they create it. This is the core design approach of Cambridge Technology Partners, which develops software systems for large organizations. CTP works on a fixed-time, fixed-budget basis and eats any overages. Its projects are rarely late or over budget, and CTP has enjoyed consistent growth (and several imitators) since 1991. In essence, Cambridge Technology Partners gets its clients to collaborate on rapid prototypes of the systems they want built. It then manages the transition from prototype to working application.

"The first thing to understand," says founding CEO Jim Sims, "is that 90 percent of the problems in software happen before you write the first line of code. Applications are late because the people building them don't know what the users really want—which means they keep changing. Or they're late because users ask for every feature they can imagine, on the theory that if they don't get it now they'll never get it. In most companies, the front-end design process alone takes six to nine months. In our Rapid Solutions Workshop we do it in three weeks. The time compression is incredible. It can be an ugly, frightening experience at first. . . . Users are laying out what they need, and the technology people want to crawl into a hole. Our people guide [the users] through what features make economic sense, what you can tie to real economic benefits. . . . I tell you, the results are amazing."

CTP is selective about the applications it will codevelop for clients. "We build only what we call 'numerator applications'—software designed to generate revenue rather than cut costs," says Sims. "They're designed to increase market share and get products to market faster. Senior executives will get very involved in those projects. And users will stay involved as the system gets designed." The first goal of the workshop is to surface the key design and implementation issues as rapidly as possible. CTP pursues this goal by relentlessly forcing the client team to build prototypes of the proposed software. "We want to see how clients behave around the software they're prototyping," says Sims. "We want to force them to prioritize and hear the reasons why they're prioritizing."

> "It's rare that companies build the apps they thought they would build."

The client team's conversations and negotiations change as the prototyping iterations mount up. "It's rare that companies end up building the apps they came in thinking they would build," Sims reports. "What usually happens is that the team discovers what they really want as they build their prototypes. . . . That clients end up surprising themselves is no surprise to us. The Rapid Solutions Workshops force them to recognize that their original priorities might not be the best ones for building the systems that add the most economic value fastest." It's an article of faith in the CTP culture that rapid prototyping shakes up the client's core assumptions.

The point is that the prototype doesn't represent the product of a methodical development path; instead, it emerges from interactions around iterations of the prototype. The solutions emerge from the evolution of the prototype, enabling clients to usefully and productively surprise themselves.

John Rheinfrank, whose industrial and interface design work with such companies as Xerox and Texas Instruments has won awards and market share, has long insisted that clients play with product prototypes in the context of customer scenarios. "If everything goes the way we thought it was going to go, then clearly we did something wrong," he says. "The purpose of getting people to play with prototypes is to capture the unexpected, not confirm the findings of some focus group. If a model or simulated environment isn't creating behaviors and choices that surprise you, then you probably are being too conservative. You probably aren't being particularly creative or innovative."

In Rheinfrank's context—and in the context of a CTP or any other model-intensive environment—issues of creativity and surprise are intimately connected with the challenge of expertise and craft. In a wonderful mini-essay, "Surprise, Craft and Creativity," the eminent Harvard psychologist Jerome Bruner deftly explains the link between the craft of modeling and the creation of surprise:

> If the creative product has about it anything unique, it is its quality of surprise. It surprises, yet is familiar, fits the shape of human experience. Whether truth or fiction, it has verisimilitude.
>
> Surprise in the creative takes three forms. One is the surprise of the fitting but unlikely, an empirical or functional surprise. "How clever to use that in that way!" Psychologists use the principle in a "multi-

*ple uses test" to find whether someone uses objects "creatively."
Empirical surprise is ingenious rather than deep. Consider formal
surprise, by contrast. Take hands of bridge. Any hand is equally
unlikely. Some are extremely interesting—all spades, for a case.
What makes such a hand interesting is not its improbability but its
relevance to a rule structure. That feature is at the heart of formal
surprise. Suppose one produces a solution to a mathematical problem that is within the formal constraints of the rule system, yet is
both shockingly new and yet obvious (once done). Almost inevitably,
such a product will have both power and beauty. The powerful simplicity of the great formulations in physics are a case in point.*

*Formal surprise is also to be found in music. Music has structure;
composers have signatures within that structure. One can simulate
Bach-like or Mozart-like music by computer. It is quite banal. Yet
both Bach and Mozart used the same sort of "programme" and
produced surprises—and not just historical ones. Their nature is
harder to characterize since music lacks anything by way of the
truth-testability of elegant scientific formulations or the consistency
tests of logic.*

*Finally, there is metaphoric surprise. Its shock value depends upon
that structured medium of language and symbols. Metaphoric surprise opens new connections in awareness, relates where relations
were not before suspected. . . . Why does [a poem] illuminate? What
does it satisfy? The answer remains obscure.*

*While our three forms of surprise have creative novelty about them,
they are almost always the fruit of disciplined craft. The proverb
about poor workmen blaming their tools is relevant. Auden comments that a poet "likes to hang around words." Good painters cannot let go; neither can the good mathematician, though his hanging
on (like Lewis Carroll's) may be full of fun. For the production of
creative surprise demands a masterful control of the medium. It is
not the product of spontaneous seizure, an act of sudden glory.
Music and mathematics give gifts to the well prepared. So, too,*

*poetry, and engineering. How curious that surprise grows in the soil
of grinding work. A woman at a dinner party is alleged to have said
to Alfred North Whitehead, "We are all philosophers, you know."
"Yes," he agreed, "but some of us spend all day at it."*

Whether their models, simulations, and prototypes are built
around functional, structural, or metaphorical surprise, the most
innovative organizations spend all day at it.

*"Surprise, Craft, and Creativity," in PLAY: ITS ROLE IN DEVELOPMENT AND EVOLU-
TION, edited by Jerome S. Bruner, Alison Jolly, and Kathy Sylva (Harmondsworth: Penguin
Books, 1976), pp. 641–642. Copyright © Jerome Bruner, Alison Jolly, Kathy Sylva, 1976. Repro-
duced by permission of Penguin Books Ltd.

PERILS OF PATHOLOGICAL PROTOTYPING

This is a chapter about failure. Instead of exploring the critical success factors that give prototypes their power, it examines the critical failure functions that assure irrelevance and impotence. Some of the most seemingly sophisticated prototypes in enterprise management are little more than voodoo dolls dressed up in quantitative clothing. They do not represent virtual realities in which meaningful tradeoffs are explored but fantasies in which hard choices are avoided. Built with scraps of business reality, they are ultimately disconnected from reality. They are expensive exercises in what Nobel laureate Irving Langmuir once described as "pathological science," in which the dishonesties and flaws of human behavior consistently distort the promise of the observations.

Some enterprise-management models are not virtual realities but fantasies.

Bigger Isn't Better

In 1969 the U.S. economy was booming. A heady, technocratic optimism prevailed. Powerful new computer technology had become available, and ambitious and quantitatively driven managers were looking for new tools with which to track and manage their fast-growing enterprises. That year, the *Harvard Business Review* published an article by George Gershefski about how his Corporate Economic Planning group at Sun Oil Company—then a $2-billion-a-year giant—built its corporate financial model, possibly "the largest and most complex corporate model yet developed."

Gershefksi's article was an account of a dazzlingly successful transformation of a complex organization. Sun Oil's bold initiative was presented as a model for corporate model building:

> A well-developed corporate financial model not only collects and stores data, but also processes and presents them in such a way that they are useful for decision making. For this reason, I think that the corporate financial model will be at the heart of future management information systems.
>
> But the corporate model has become a powerful tool in its own right. It is extremely valuable for comparing and evaluating alternative courses of action that a company may take. . . .The model enables management to react quickly to events and to revise estimates of income and other aspects of performance. . . .
>
> In view of the rapid changes taking place in modern life and the increasing complexity of carrying on a business, the need for structured information is more vital than ever. The financial model, together with advanced computer technology and information systems, can provide management with a way to tackle these challenges. ("Building a Corporate Financial Model," July–August 1969, 61–72)

These conclusions have stood the test of time, but Sun Oil's corporate financial model did not. Within two years of the article's publication, Sun Oil disbanded its strategic modeling apparatus. It turned out that the model had virtually no impact on management of the enterprise. One observer described Sun Oil's experience as an abject failure. "For all practical purposes," noted strategic-planning consultant Thomas Naylor in 1981, "the model was really never even used by Sun's management."

In hindsight, it's clear that Sun Oil's bold investment in digital simulation was destined for dismissal from the outset. Its overweening ambition was exceeded only by its technocratic arrogance. Bigger was

The Sun Oil model had virtually no impact on management of the enterprise.

better. The more variables the merrier. According to Gershefski, a working version of the model required thirteen man-years to develop and another ten man-years to create management aware-ness. The total number of equations necessary to run the model was "some 2,000 . . . with 1,500 inputs and 5,200 output items." In practice, Sun Oil's modeling methodology grotesquely twisted Alfred North Whitehead's famous admonition to "Seek simplicity and distrust it" into "Trust complexity as long as it's been compre-hensively modeled." This distortion virtually guaranteed that Sun's model couldn't be used by a human organization.

The simulation was designed to correspond faithfully to the physical and financial processes of the firm. Simulation in this con-text meant imitation. It was as if a mapmaker had decided to pro-duce a map as big as the city it depicted, with a perfect 1:1 corre-spondence. Who would actually use such a map? If you could manage Sun Oil's simulation, you could manage Sun Oil. The hopeful technocratic assumption of Sun Oil's modelers was that their simulation would seamlessly blend into the realities of Sun Oil's operations. But that could never be.

Why? There are a number of excellent technical reasons. But the best reasons have to do with how people and organiza-tions behave. The most shocking aspect of this story is the disconnect between this

It was as if a mapmaker had decided to produce a map as big as the city it depicted.

sophisticated simulation and the everyday realities of managerial life. The point isn't that the model builders ignored Sun Oil's managers, but that it was never quite clear what the simulation would actually do for them.

As Gershefski noted at the time, "Early in the model's development an attempt was made to define its potential uses. This was extremely difficult to do, since few persons knew what a corporate model was or how it could be of value. . . . Moreover, experience had shown that the model was a very difficult concept to explain and the group decided that the best demonstration of its worth would be the completed system itself."

So the organization as a whole didn't understand the value of the simulation, and influential managers never explored a prototype of the simulation before it became a working model. No one addressed how the simulation would be integrated into the political, cultural, and managerial values of the firm. Most important, the enterprise had no way of measuring when Sun Oil might break even on its twenty-three-man-year investment. There was no credible analysis of how or when the benefits of Sun Oil's corporate financial model would outweigh its costs.

Nearly three decades later, Gershefski acknowledges that "there was no need or demand for what the model could do; we had a screwdriver and we were running around looking for a screw." Thus technical tour de force—even if it could have proven useful—had all along been decoupled from the operations of the firm. There were no incentives for using it; there were no penalties for not using it; and its value was neither intuitively nor analytically obvious to managers at Sun Oil.

> "We had a screwdriver and we were running around looking for a screw."

But Sun Oil's experience with its strategic simulation software is notable because it was not an aberration; as a literature search confirms, it represented a corporate norm. A statistician observed in a 1973 *Journal of Business Policy* survey: "In a large number of firms, model developments have been severely reduced or stopped entirely. Many of the planning models have not been implemented or are used only on infrequent occasions."

Failures of Use

Models of the Sun Oil variety tend to collapse of their own ambition and flawed complexity. Ever since the advent of operations research and systems dynamics after World War II, rational managers have been trained to believe that modeling the right problem in the right way will lead to the right solution. Unfortunately, this precept has not been borne out. Too often the actual costs of designing, building, implementing, and managing models outweigh their purported benefits. Market forces reject the value proposition. This is the real legacy of the postwar business-simulation experience. The supreme irony is that the technical quality of the model rarely determines its impact on the enterprise.

> The technical quality of the model rarely determines its impact on the enterprise.

The proper question is not "How will this model or simulation solve the problem?" but rather "How will this simulation or model *be used* to solve the problem?" As Gershefksi ruefully acknowledges, "The model itself is nothing." There has been too much focus on the quality of the model and not enough on how using the model will change the organization's behavior.

Models don't solve business problems, any more than mathematics solves equations. How models are used determines whether and how problems are solved. And questions of use are always—without exception—matters of corporate culture, individual behavior, and the political economy of the firm, rather than of rational analysis or technical expertise. Questions of use are really questions about who chooses and uses the model and why; they're questions about when and where it's appropriate to apply simulation technologies or prototypes. Use is contingent on factors that are often completely unrelated to the original intentions of the model builders. Just ask Gershefski, or any one of the thousands of corporate model builders whose expensive failures litter the business landscape.

Talking about simulations and prototypes without regard to how they will be used is like talking about the medical effectiveness of chemotherapy independent of dosage: too much can kill, too little has no therapeutic impact. Even the proper dose may induce undesirable side effects. This is why successful medicine, like successful modeling, is more craft than science. It's difficult to predict how well individual patients will respond to complex treatments. It's even harder to predict how individual organizations will respond to complex models and simulations.

Models tend to be far more consistent than the people who use them; they're designed to be. By focusing on the behavior of the models rather than that of the people who use them, organizations practice the most destructive sort of self-deception. The novelist George Eliot beautifully captures the human challenge of managing complexity in the context of simulation in *Felix Holt:*

> *Fancy what a game at chess would be if all the chessmen had passions and intellects, more or less small and cunning: if you were not only uncertain about your adversary's men but a little uncertain about your own . . . if your pawns, hating your guts because they are pawns, could make away from their appointed posts that you might get a checkmate on a sudden. You might be . . . beaten by your own pawns. You would be especially likely to be beaten if you depended arrogantly on your mathematical imagination and regarded your passionate pieces with contempt.*

A model decoupled from its reality—the passions and politics it can arouse—is little more than an intellectual exercise. Intellectual exercises can be extraordinarily valuable. They can stimulate useful insights and new ideas. But a model that exercises the intellect may not be closely coupled to the task of managing value in the firm. Few believe that chess grand masters—Fischer, Karpov, Kasaparov, Deep Blue—would necessarily make great military generals or superior business executives. Strategic brilliance in

A model decoupled from politics is little more than an intellectual exercise.

chess does not assure strategic brilliance in business or war. Nor do brilliant models assure brilliant management; the brilliant *use* of models, however, can.

Decouplings and Disconnects

Corporate cultures and institutional infrastructures that decouple smart models from smart management can be found in every kind of company that departmentalizes and delegates to manage complexity.

Design versus manufacturing

At General Motors in the 1930s, manufacturing engineers often cursed at the beautifully sculpted clay models from Harley Earl's design group. Yes, Earl's "clays" were regarded as bold and sexy, but the company's top engineers considered many of the most elegant prototypes utterly unmanufacturable. The designers themselves acknowledged that their clays were crafted with little thought of assembly lines. Details like fins and grills could be mass-manufactured, but the body was a different matter. Earl's clays often generated more contention than innovation. The resulting arguments could add months to a new automobile's planned schedule and millions of dollars to its budget.

GM's prototyping pathology was hardly unique; Ford and Chrysler body stylists also carved clays that their manufacturing engineers couldn't build. Top management typically justified the separation of designers from engineers to preserve the artistic purity of their designs. At the same time, this compartmentalization preserved the power of the manufacturing engineers, who could argue—with total justification—that designers ignored the realities of metallurgy, dies, molds, and mass-production techniques. Design historian Stephen Bayley describes one of Harley Earl's notable design innovations: "The engineers at the Libbey-Owens-Ford glass company were so confounded by the precise shape of the curved windshield that Earl had instructed them to fabricate that they failed to find a mathematical formula to define the actual path [he] wanted the glass to follow. The numbers to which Earl paid homage were not those know to Pascal, but rather those tested by Gallup."

The clays proved extraordinarily useful media for visualization, but they seldom promoted collaboration media because they weren't built with collaboration in mind. Indeed, the clays were the property and preserve of the stylists, not the engineers. The nature of the medium also itself militated against easy changes. Modifying a clay on the margins was relatively easy: scrape off a millimeter here, round off an edge there. But making a significant change was akin to sculpting a brand-new David. Meanwhile, manufacturing's prototypes were organizationally beyond the reach of the stylists. For decades, designers and engineers each built their own prototypes. From the 1940s right through the 1980s, "concept car" was an accurate Motown euphemism for breakthrough automobile designs that could never be built.

> **For decades, designers and engineers each built their own prototypes.**

For decades the U.S. automobile industry had a serious disconnect between the cars its most creative stylists designed and those its factories actually built. Were the clays—the most beautiful and expensive voodoo models in American industry—without value to the enterprise? Of course not; clay was the birth medium for many smart styling innovations. Did the clays add value when measured against the time, costs, and conflicts incurred? The answer to that question is less clear.

Finance versus marketing

There is an interesting commercial coda to this story. During the 1980s, when General Motors hemorrhaged both money and market share, the firm sought to reuse many of its platforms, subsystems, and components in an effort to leverage economies of scale and dramatically cut costs. Manufacturing's prototypes embodied the engineering theme of reusability. On a pure cost-accounting basis, this new design paradigm was successful. Unfortunately for GM, what auto reviewers and the automobile-buying public saw were fleets of look-alike/drive-alike cars with no distinctive design features to capture the imagination.

The prototypes of GM's so-called J-cars did a superb job of adhering to finance's cost imperatives. However, these prototypes were decoupled from the needs of GM's sales and marketing organizations, not to mention its dealer network. The result was a fleet of cost-effectively built cars that sold significantly below GM's expectations. Finance-driven prototypes proved as disconnected from the marketplace as designer-driven clays had been from manufacturing and engineering.

R&D versus production

The same decouplings and disconnects can also afflict the design and development of a product as mundane as disposable diapers. A huge consumer-products company spent tens of millions building a facility that did nothing but produce prototypes of disposable diapers. The goal was to produce enough prototype diapers to adequately test proposed improvements and innovations. Having cut back on the number of its production lines to better manage costs and optimize manufacturing throughput, the company had decided that it could not afford to test proposed enhancements on its regular diaper-production lines. Indeed, the diaper production managers were explicitly rewarded for their ability to eliminate downtimes and maximize production levels. Intriguingly, they were not rewarded for introducing or testing new diaper developments.

The prototyping facility represented a corporate initiative in support of R&D. But there was little direct consultation between it and the production facilities. The production lines required proprietary equipment that blended sophisticated mechanical engineering with chemical-engineering technologies; most of the prototype diapers were handmade. It was thus completely unclear whether any of the handmade prototypes were manufacturable.

The company offered no explicit incentives or rewards for successfully transferring ideas from prototype to the actual manufacturing process, nor for encouraging the manufacturing and production engineers to build tighter relationships with the prototypers. In effect, the diaper prototypes resembled Detroit's clays—interesting and provocative designs decoupled from the actual production process. For all intents and purposes, the prototype facility represented an innovation ghetto.

> **The prototype facility represented an innovation ghetto.**

2-D versus 3-D

Decoupling isn't always a by-product of organizational design. Don Norman, an astute observer of design management, reports that industrial designers at Apple Computer initially had trouble translating their sketches into the CAD programs linked to manufacturing software and that their designs thus had minimal influence on the actual prototyping of the machines. As the CAD software improved, Norman notes, the designers could comfortably sketch their best ideas directly into Apple's production infrastructure. The ability to sketch and the ability to digitally design were no longer technically disconnected. The result, Norman asserts, was better designs faster.

Manufacturing firms worldwide have been transitioning their computer-aided design platforms from two-dimensional renderings to three-dimensional volumes. The reason is straightforward: two-dimensional representations of three-dimensional objects can introduce too many ambiguities and implementation issues for error-free manufacture. As the costs of three-dimensional "volumetric" and "solid modeling" have dropped, firms worldwide are transferring their engineering design to software on which all three dimensions are integrated from the object's moment of conception.

Tight coupling

The tighter the coupling between a prototype and its real-world use, the more power and influence it is likely to possess. Perhaps the most disciplined proponent of this approach is Intel, the world's dominant microprocessor manufacturer. As Intel chairman Gordon Moore puts it: "With a product as complex as semiconductors, it is a tremendous advantage to have a production line that can be used as a base for perturbations, introducing bypasses, adding steps, and so forth. Locating development and manufacturing together allows Intel to explore variations on its technology very efficiently."

Intel is among the world's most sophisticated users of modeling and simulation technologies in part because of this rigorous practice of linking exploration with implementation. Moore elaborates:

> *Intel operates on the . . . principle of minimum information: one guesses what the answer to a problem is and goes as far as one can in a heuristic way. If this does not solve the problem, one goes back and learns enough to try something else. Thus, rather than mount research efforts aimed at truly understanding problems and producing publishable technological solutions, Intel tries to get by with as little information as possible. To date, this approach has proved an effective means of moving technology along fairly rapidly.*

This principle of minimum information does not signify a belief in willful ignorance. Instead, Intel is promoting as a cultural value collaboration around a practical model of the problem at hand rather than a theoretical understanding of it. Because the company wants its innovation culture to focus on practical solutions to practical problems, its prototypes must be tightly coupled to the design or production problem in order to adhere to the principle of minimum information. Intel does not want to take the chance that the cost of prototyping will ever outweigh its benefits.

"Intel tries to get by with as little information as possible."

To underscore the importance of tight coupling, Moore frankly acknowledges that it comes at a serious cost: Intel does a relatively poor job of accommodating dramatic change. "Whenever Intel has explored a totally new technology, such as bubble memory, it has set it up as a separate organization," explains Moore. Presumably the new organization adopts a comparable ethic of tight coupling its prototypes to production.

Black Boxes

If the prototype is flawed, however, tight coupling can choke the profit right out of a product or process. Certainly, the culture of modeling and simulation at Kidder, Peabody was one of Wall Street's unhealthiest. Kidder's prototyping pathology tightly coupled "black-box" models that injected flawed assumptions into its government-bond trading business.

The story is a cautionary tale about how the cultural flaws of an organization can grotesquely amplify the technical flaws of a model. Kidder's misuse and abuse of financial-simulation software on an institutional level is what ruined its credibility and its finances. Deliberate fraud may have been a factor, but Kidder's unhealthy cultural dependence on "black-box" trading simulations ultimately led to its self-immolation. How so?

In 1991, Kidder hired Joseph Jett to arbitrage treasury bonds and STRIPS (separate trading of registered interest and principal of securities, i.e., bonds stripped of their coupon payments). Such arbitrage is theoretically a riskless transaction and would thus not need to be tracked by Kidder's standard market and credit risk management systems. The firm relied on a computerized expert system that allowed traders to model and simulate their trades in accordance with software rules about valuing such transactions in the bond market. The

The cultural flaws of an organization can grotesquely amplify the technical flaws of a model.

software also automatically updated the firm's inventory, position, and profit-and-loss (P&L) statement. In keeping with market conventions, the system valued the STRIPS lower than their associated bonds. This difference was reflected in the firm's P&L statement, which was also the basis for assessing trader bonuses. By entering into forward transactions on the synthetic STRIPS, Jett was able to defer when the actual losses were recognized on the P&L statement by taking up still larger positions in STRIPS and then digitally reconstituting synthetic STRIPS already in the system.

In 1993 Jett enjoyed STRIPS profits in excess of $150 million; he received a $12-million bonus and the chairman's "Man of the Year" award. By March 1994, when Jett's positions included $47 billion worth of STRIPS and $42 billion worth of reconstituted STRIPS, Kidder management decided to figure out Jett's secret. A month later, the firm announced that Jett had falsely inflated his profits in excess of $350 million. He was fired and sued for fraud. Charges and countercharges over whether Jett's actions represented deliberate fraud or a ruthlessly clever exploitation of flaws in the software snaked through the courts, the Securities and Exchange Commission, and the bond-trading community.

The indisputable fact, however, is that no one at Kidder rigorously reviewed or replicated Jett's trading strategy. Kidder's culture treated success as an untouchable asset, and a successful trader's tactics as a black box. Trading strategies and simulations were not peer-reviewed or subjected to independent analysis by the accounting or control functions. Traders effectively "owned" their own black-box models. To the winners went the bonuses; the losers left, and who really cared about scrutinizing the middling traders?

According to *Institutional Investor*, "Jett had been trained . . . to accept information from the computer screen and to act on it. If the computer got it wrong, he was going to get it wrong; and if the computer got it wrong in his favor, he would see no reason to ask questions, even if things got silly." Jett's software simulations invariably told him to take bigger positions at higher stakes. He executed those trades, profitably, for two years. During that entire time, Jett's tactics were never subjected to meaningful scrutiny. No one ever audited the model or ran simulation stress tests on the systems software, even though it was directly linked to the firm's P&L statement!

If Kidder had developed a prototyping culture in which trading strategies could be safely peer-reviewed in simulation or digitally stress-tested, there is little likelihood that a systems failure of this magnitude could have occurred. But Kidder's prototyping pathology effectively guaranteed that when—not if—a trader or trading desk crafted a highly sophisticated but ultimately destructive strategy, it would inflict the maximum damage on the enterprise. Kidder's culture downplayed simulation as a medium for risk management. Instead, simulations were exclusively tools for optimizing the individual trader.

Peer review in simulation or digital stress-testing could have prevented the Kidder, Peabody systems failure.

The lesson? The coupling of a flawed technology with a crippled culture can create truly destructive realities. Kidder's culture of simulation did not ignorantly stumble into disaster; it deliberately and consistently chose ignorance over self-knowledge.

"Biggerism" and "Feature Creep"

Perversely, the very effectiveness of prototypes as innovation platforms can tempt organizations to make them too ambitious. "The moment you successfully demonstrate a concept can work," says IDEO's David Kelley, "you have people trying to add features to make the prototype even better. It becomes a Christmas tree." Design success can breed feature excess.

In software development, for example, numerous studies affirm that "feature creep" is one of the most common sources of cost and scheduling overruns. Even at companies with reputations for rigorous development practices, "featuritis" can infect the prototypes. "Product development at Microsoft is 'feature driven,'" observe Michael Cusumano and Richard Selby in their essay "Microsoft's

Weaknesses in Software Development." "This approach tends to result in *products that contain more features than users really need* [emphasis in original]."

Not only do these added features frequently disrupt delivery schedules and introduce bugs, they undermine the integrity of the product's design. According to Cusumano and Selby, "This preoccupation with features encourages projects to *underemphasize the importance of the underlying product architecture* [emphasis in original]."

In other words, Microsoft's prototyping culture makes it too easy for marginal features to be introduced into products. The unhappy result, "bloatware," has added features detracting from rather than adding to perceived product quality.

At Xerox's Palo Alto Research Center, engineers grimly joked about the phenomenon of "biggerism": the desire to add all those wonderful features left out of the initial prototype because of time and budget constraints. As Fred Brooks asserted in *The Mythical Man Month*, "The second is the most dangerous system a man ever designs."

While it's tempting to exploit the flexibility of prototypes to improvise and test new features and innovations, this comes at a price. At Boeing, the chief engineer of the 777 reportedly hung a sign on his desk to discourage feature creep and biggerism in the plane's system software: "No! (What part of this don't you understand?)" The most important role a prototype can play is often to creatively limit the scope of a proposed innovation rather than to broaden functionality beyond a point of diminishing returns. Once again, the prototype becomes the medium for managing the critical trade-offs in design and development.

Perils of Prediction

Of all the organizational pathologies of prototyping, none is more pernicious than those associated with predicting the future.

It's completely understandable why managers, investors, and entrepreneurs crave tools and media with which to forecast their futures. Corporations worldwide invest tens of billions of dollars annually in people and technologies that promise actionable insight into tomorrow. The results are mixed at best.

Certain physical, biological, and economic phenomena do yield to predictive quantitative modeling and simulation, but most organizational and institutional behaviors do not. Thus efforts to run for-profit enterprises by depending on the predictive powers of econometric models or operations-research simulations or systems-dynamics constructs are no more successful than long-range weather forecasts.

Most organizational and institutional behaviors do not yield to predictive quantitative modeling.

The point is not merely to puncture the belief that "better" science and technology will make the future more predictable but to emphasize that such futurology undermines the genuine potential value of future-oriented models. As they say at Xerox's Palo Alto Research Center, where the Ethernet and object-oriented graphical user interfaces were first built, "the best way to predict the future is to invent it."

Truly effective models and simulations aim not to predict the future but to envision possible futures that can be managed successfully. This is the central message of the scenario-planning work so superbly pioneered by Arie de Geus and Pierre Wack at Royal Dutch/Shell. Under their leadership, Shell rejected planning as prediction in favor of planning as learning. The result was an organizational culture able to play comfortably with the manifold possible profiles of alternate tomorrows. Shell's scenarios deliberately sacrificed false certainty for flexibility. Henry Mintzberg's repudiation of formal strategic planning and Amar Bhide's clever reformulation of "strategy as hustle" reinforce this core belief that organizational learning is a far better investment than organizational fortune-telling. The safest prediction about the future is that it will be impossible to predict.

Recognizing and Avoiding Critical Failure Functions

Prototyping pathologies are easier to identify than to avoid. But they can be managed only if they are honestly diagnosed and confronted. The presence of a given "critical failure function" does not in itself guarantee failure. The cruel irony is that a model or simulation without technical, scientific, or financial validity may still garner enthusiastic support. Astrology and phrenology are only two examples of technically flawed models that have enjoyed sustained acceptance.

Costs that outweigh benefits

Cost is a variable that transcends mere time and money: there are also coordination costs, opportunity costs, political costs, and behavioral costs. Too often, models are assessed in terms of the problems they will solve or the opportunities they will create, rather than the multiplicity of costs they will incur. Who assesses whether a model creates more problems than it actually solves? Might the cure be worse than the disease? These questions plague every effort to manage reality by modeling it. Using models as they are intended always creates, along with unintentional consequences, unwelcome costs.

> **Using models as they are intended always creates unwelcome costs.**

Unless it has been thoroughly thought through in advance how the benefits of the model will comfortably exceed its costs, both qualitative and quantitative, the Sun Oil fiasco will repeat itself. The costs and risks may turn out to be unrealistic—but those fears are themselves costs.

Decoupling from the point of economic impact

As we have seen, a beautifully engineered product prototype may be impossible to manufacture and thus effectively useless.

Conversely, a few differential equations scribbled on a whiteboard might translate into process software capable of managing a sophisticated chemical-engineering plant. In other words, a model may perfectly resemble reality but be utterly decoupled from it; or it may appear abstract but have immediate concrete implications. The impact of a model is a function of how tightly coupled it is to the ultimate marketplace.

Unless a model is tightly coupled to its real-world impact—or unless a clear path to impact has been mapped out—its economic value is in question. As we have seen repeatedly, effective models do more than change awareness; they change behavior. They change the choices people make. Models decoupled from key links in the value chain, like Harley Earl's unmanufacturable clays, lose power and influence. Models and simulations tightly coupled to mission-critical processes, as at Intel, amplify productivity. Organizations that can't cost-effectively couple models to reality waste time, money, and effort.

Lack of appeal

Like people, models can have intrinsic appeal. By the same token, competent and potentially valuable models can also be ignored because no one cares to work with them. The notion that a good model will be used simply because it works is wishful nonsense. Perfectly good models—like perfectly good employees—may be underutilized investments that create only marginal value for the enterprise. Models often have to be marketed aggressively. If Sun Oil's top management had mandated use of its Corporate Economic Planning group's new computer simulation for divisional budgeting, the story might well have turned out differently. Had Sun's treasury taken a shine to the software, for instance, the system might have been accepted and adopted. Many of GM's clay models exerted influence because top management wanted their features reflected in the finished cars. Charisma confers power.

Charisma confers power.

But when models are not mandated, lack intrinsic appeal, or demonstrate a blurry value proposition, they are reduced to the status of drones. Too many organizations produce too many drone models and simulations, which in turn breed a cultural pathology of ineffectuality.

Provocation of hostilities

Some models and simulations generate immediate hostility, usually because their use threatens to undermine the organizational status quo. To quote Cicero, the operative question is, "Cui bono?" Who benefits? Politics and culture can easily trump economics and value. As former Sun Microsystems designer Jakob Nielsen points out, fundamental differences between the choosers and the users of prototypes reflect issues of power and control. The politics of prototyping will be explored at greater length and depth in Chapter 7. Suffice it to say here that organizations in which the design and use of models is excessively influenced by politics almost always generate pathological prototypes.

Inaccessiblity and incomprehensibility

No one at Kidder, Peabody understood what Joseph Jett was doing—not his colleagues, not his bosses, and certainly not the internal auditors—until it was too late. Kidder's culture actively promoted the use of trading simulations that were inaccessible to key constituencies within the firm. Sun Oil's corporate financial model was impenetrable by virtue of its cumbersome complexity. Top management couldn't be bothered to plow through the complexity to get at the core value. In both cases, the inaccessibility of the model played a direct role in its failure.

Because quantitative complexities can intimidate managers who lack the requisite training to interact with them, electrical-engineering, chemical-engineering, and manufacturing models often end up ghettoized and decoupled from the mission-critical value-creation processes of the firm. Instead of placing their trust in models they don't understand, even well-educated managers place their faith in the people who present the models. If they don't trust the people, the models don't stand a chance.

> **Instead of trusting models they don't understand, managers trust the people who present the models.**

If models can be made more accessible through techniques of simplification and visualization, without undermining their fundamental validity, they stand a better chance of being more tightly coupled across functions and processes. This is why so many organizations invest more time and effort in improving "user-interface" design.

Failure to create a constituency

Models that do not evolve interactively with the managers, customers, and suppliers who use them grow stale. Such models become frozen in time, more like snapshots than cinema. When models are untouchable—when it would be unacceptable, say, for a manufacturing engineer to alter a clay or for a fellow trader to tweak Jett's arbitrage software—they preclude new investment in them by players who might be able to reap more value. Truly effective models aren't just effectively used; they are effectively cultivated to become even more valuable through use and to create vibrant markets around themselves.

> **Models that do not evolve interactively grow stale.**

Emphasis on prediction over projection

Failed predictions breed contempt for the models that generate them. By contrast, organizations that use modeling media to learn more about themselves and their markets invariably report that their investment is worth the cost. This is not to say that no effort is made to forecast and predict, only that predictions are bracketed by probabilities that place them in manageable contexts. Organizations that try too hard to eradicate the uncertainties of the future often have trouble dealing with them when they—inevitably—arise. Using models to gain insight into the future, and even to influence or create it, is healthy. Relying on these tools to predict the future is about as sensible as eating prime rib with chopsticks.

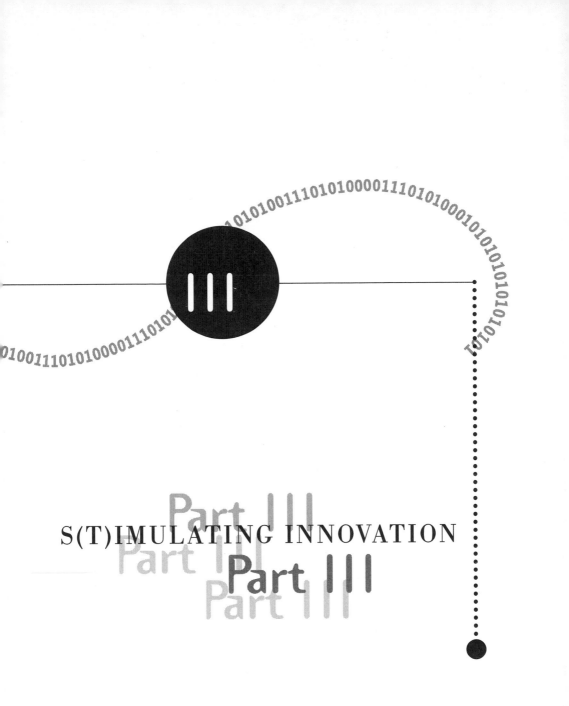

S(T)IMULATING INNOVATION

Part III

S(T)IMULATING INTERVENTIONS

One cannot teach a man anything. One can only
enable him to learn from within himself.
Galileo

During a business simulation game, a senior manager—
who happened to be a Holocaust survivor—made a bold and unexpected move. The gambit was neither unethical nor particularly
unfair. In the context of this game, however, it was ruthlessly self-
serving.

"So," remarked a colleague, "is that how you survived the Holocaust?"

Visibly stunned, the stricken manager remained subdued for the
rest of the session. Whether snide or sincere, a single comment
transformed this game into something else. A seemingly benign
business simulation had trespassed onto the most intimate and
painful of personal realities. Was this remark the start of a serious
conversation? Or an act of insufferable insensitivity?

There is no answer. This vignette is not about rudeness or
candor; it is about behavior. It highlights the fundamental misun-
derstanding about the role that prototypes play in organizations.
There is a belief—zealously championed by engineers, economists,

operations researchers, strategists, and systems designers—that the value of models and simulations is best found in the insights and information they help create.

That is eminently rational but dangerously untrue because organizations and the people who run them are not eminently rational. To treat models and simulations as "information engines" is a mistake. Management decisions are more than processed data; organizational behavior is not some irritating by-product of dispassionate analysis. Leaders and their management teams optimize and "satisfice" using different criteria. So, unsurprisingly, they respond differently to prototypes chosen to manage risk and opportunity. Who knows what innocent comments an individual might make around a simulation? Who knows what conflicts a prototype might provoke?

Precisely because organizations are communities of people and not aggregations of data, the real power of prototypes comes less from the prototypes' acknowledged abilities to model reality than their implicit threats to change behavior. To become truly engaged in a prototype is to create a new relationship with the self and with others.

Prototypes are best viewed through the prisms of human behavior than through the lenses of information technology. Why? Because human interactions matter more than information. Changing behavior matters as much as changing minds.

Power Trips

Policy and custom expressly forbid the president of the United States from active participation in decision making during national-security war games. The secretary of state plays, the joint chiefs play, the director of the Central Intelligence Agency plays, only the president does not. The U.S. national-security establishment has decreed that no one should know how the president might react to speculative scenarios. Presidential advisers in real national-security emergencies should not be influenced by prior knowledge of how the president responded to a simulated crisis. Nor should potential enemies of the United States. The president should in effect be above simulated frays. The president attends to see how his advisers respond, not so that his advisers can see how he responds. Indeed, the president's participation would undermine rather than enhance the value of the exercise.

The Pentagon discovered during 1960s-era war-game exercises that pitting officers of different ranks against one another didn't work. Well-meaning military innovators had believed that mixed-rank exercises would enhance career development and encourage nonhierarchical communication, but the actual result was rivalry and recrimination. "The Pentagon found out that you don't play generals against colonels," recalls defense-simulation designer Clark Abt. "You play peer to peer."

John Rheinfrank, a noted industrial designer who has developed successful products for firms such as Xerox and the United States Army, tries to avoid delivering the shock of the new: Rheinfrank is careful about how he presents "breakthrough" prototypes to his more culturally conservative clients. "Pacing the prototypes turns out to be more important than we had initially thought," he says. "You don't want to go a prototype too far. . . . You want to surprise clients, not insult them."

These examples all illustrate the point that managing power and influence is more important than managing knowledge and information. And each example raises fundamental questions. Has the belief that the president should not participate in war games been scientifically tested and verified over the course of time? Has the United States gained a demonstrable competitive advantage as a result of this practice? Should Fortune 500 chief executives take their cue from the White House and abstain from business simulations to encourage a more open exchange of ideas among their subordinates? Should the military's disastrous experience with mixed-rank games be respected? Or is it a warning sign of a dysfunctional organization? Do Rheinfrank's carefully paced prototypes bespeak his clients' conservatism or his own? Does his incremental approach prepare clients to accept breakthroughs, or does it merely reinforce their incremental innovation cultures?

In every one of these examples, the model is a battlefield on which power relationships are enacted. Managing the prototype means managing these relationships. Thus prototypes aren't essential merely for designing products and services; they're absolutely essential to designing the relationships that build the products and services. Successful industrial and simulations designers rely on key heuristics and rules of thumb to constructively manage these relationships.

> **The model is a battlefield on which power relationships are enacted.**

Pyrrhic Victories

What you reward you get more of; what you punish, you get less of. Incentives matter. Knowing the rules of the incentives game matters. Princeton mathematical economist Martin Shubik, a pioneering game theorist who has analyzed both military and business simulations, observes that all models have two sets of rules: the rules of the model itself and the rules of the larger world it inhabits.

During the mixed-rank Pentagon war games, for example, the rules of winning the simulated battles were clear. But so were the rules of the larger organization: that it's a bad idea for colonels to skillfully outmaneuver two-star generals in front of their peers. Much of the difficulty that prototypes encounter is that their rules conflict with those of the larger world. Some victories are pyrrhic.

One popular misconception about simulations and models is that they are safe: that it's permissible to take risks in a virtual world precisely because it is not the real world. This is emphatically not the case. From the Pentagon to Fortune 500 global giants to Silicon

Valley start-ups, organizational cultures punish game players who don't play by the rules. Beating the boss is usually unacceptable. Managers tend to keep a close eye on colleagues who appear too clever by half. In other words, people are sometimes punished for how well (or how poorly) they play the simulation game.

All models have two sets of rules: their own and those of the larger world.

"You can design games and simulations that get people to display all manner of pathological behavior," says Shubik, who invented the wildly controversial "Dollar Auction" game in which individuals sometimes bid more than twenty dollars for a single dollar bill. "However, aligning the incentives of the game with the incentives of the organization is very difficult. . . . The culture of a game is going to be different than the culture of an organization."

For example, would a design team be rewarded in the real world for figuring out how to creatively eliminate their subassembly from the prototype and thus from the product? Or would they discover that they'd made themselves redundant? Does prototyping to make a product easier to maintain win compliments or hostility from the service department? Does manufacturing's budget help pay for the design of components that are easier to assemble? Or is every group responsible only for its own function?

Recall from Chapter 4 that Chrysler won over its suppliers by letting them keep half of any cost savings their initiatives generated. Chrysler and its suppliers thus both reaped the benefits of collaborative cost reduction. The incentives for improving prototypes were congruent with the incentives for improving the car. The demands of a firm's internal market rarely mirror the demands of the external marketplace. Prototypes are where the gaps between those markets get negotiated. Do the incentives help narrow the gap or further widen it? How well are those incentives reflected in the economics, culture, or internal politics of the firm?

"One of the most difficult problems we face as designers," says IDEO's David Kelley, "are clients who haven't figured out how to integrate the process of developing a new product with the way they actually make money selling it. . . . I've lost track of the number of times we've spent a lot of time making a product look really cool only to find out that the client's sales force doesn't know how to sell cool. I've had to become a lot smarter about getting our clients to better balance their [own] needs with their customers' needs." The prototypes, Kelley agrees, are an essential part of that process.

> "We've made a product look cool only to find out that the sales force can't sell cool."

Cui Bono?

The crucial question to ask before building a prototype isn't, What is this model for? or What are we really trying to do? It's the classic question *Cui bono?* Who benefits?

Who benefits from building the prototype? Who benefits from using it? If it does everything it should, who gains power and influence? If the simulation runs flawlessly, who suffers? The notion that models are neutral artifacts in the service of knowledge creation is naive. There is nothing innocent about a chief financial officer's spreadsheet iterations during a corporate budget battle. There is nothing objective about presenting a bootleg prototype codesigned with the company's biggest customer to the new-product committee for funding. Running simulations and crafting prototypes are political acts that can either threaten or reinforce the existing power structures.

The truly insightful model is a truly inciteful model. The most powerful simulations and prototypes create winners and losers, even more readily than they create innovative ideas. The more compelling the model, the more insistent the question, Who benefits?

The prevailing piety is that the beneficiaries of modeling and prototyping should be a firm's customers. This is rarely more than a charming fiction. For example, IDEO tried gently but firmly to persuade a Fortune 500 medical-technology client to include a customer or two on their joint product-design team. Just as firmly, the client insisted that it didn't want any customers involved. The reason? The firm feared that involving only one or two customers would skew the design in ways its other customers might find unappealing or alienate those whose participation wasn't solicited. The company insisted that the design team could do a fairer, more objective job without customer involvement. Prototypes weren't shown to potential buyers until most of the critical design decisions had been locked in. In effect, the firm argued that the best way to prototype a product for its best customers was to explicitly exclude them.

> **The more compelling the model, the more insistent the question, Who benefits?**

General Motors' experience with the 1991 prototype of its Chevrolet Caprice is similar. As the *Washington Post* reported, "The design staff . . . had a reputation for being resentful of marketers, engineers and other corporate outsiders who had the gumption to tell them what was and was not attractive." A focus group strongly objected to the proposed redesign of the Caprice exterior. GM design vice president Charles M. Jordan rejected the focus group's reaction. "We decided not to do anything about it," he said. "We believed in the design. . . . All the car guys liked the design." Caprice sales in 1991 were half as high as expected, and in 1995 the model was discontinued.

By contrast, consider the 1987 hit movie *Fatal Attraction*. In the film's original ending, the villainess committed suicide off-screen. Test screenings—cinematic prototypes—found that audiences were displeased. The audience reaction was taken seriously. Director Adrian Lyne reshot an ending that offered suspense, blood, and retribution. Movie critics hated it, but audiences around the country cheered when the villainess was shot dead by the wronged wife.

The message here is not the pallid platitude about the importance of listening to customers. It's the importance of deciding up front who models and simulations are for. To say that prototyping is done for customer benefit when customer participation is tacitly ignored or explicitly rejected is a lie. Models built on lies lack credibility in the minds of both their creators and their users. Like any counterfeit currency, they corrupt fair exchange. Such dishonesty also breeds discord among internal factions with their own competing agendas for the prototype.

Former Procter & Gamble chairman Edwin Artzt recounts how the firm abandoned its tradition of product-development secrecy to find a supplier to solve a critical materials problem with its profitable Pampers line. After a contentious internal debate, P&G invited materials suppliers in to see how the company made diapers. One supplier commented that its equipment to mass-manufacture golf balls could be modified to mass-produce disposable diapers: the machines that wound rubber strands around golf ball cores could insert rubber strands in the leg openings of disposable diapers. This modification ended the perennial consumer complaint that the diapers crept up the baby's legs and allowed leaking. Sacrificing secrecy resulted in better relationships with suppliers, far lower production costs, and a demonstrably better product. Cui bono?

> **Equipment used to make golf balls could be modified to make disposable diapers.**

In all the preceding examples, the difficulty was less solving the technical problem at hand than deciding who wins and who loses as a direct result of how the model is managed. It makes sense for manufacturers to build prototypes to test the manufacturability of a product and for designers to collaborate with manufacturers to explore ways to reduce the number of components. But this is when "Who benefits?" creates unavoidable conflicts. Understandable tensions arise when design and manufacturing debate just how manufacturable a proposed product may be. Legitimate disagreements emerge about who should benefit.

The role of prototypes should be to make these trade-offs more explicit rather than more obscure. Problems arise when organizations are disingenuous or dishonest about the beneficiary of the prototyping process. Whenever an organization wants to prototype, it must ask itself who stands to gain and lose the most. A prototype can have a bigger impact on the organization than on the problem.

Savvy design and engineering consultants and systems modelers predicate new consultations on the premise that understanding who benefits is essential to creating models that productively influence client behavior. Without that initial understanding, the cure can be worse than the disease. Whenever the answer to "Cui bono?" is, "We're not sure" or "We'll let you know," experience dictates that the model is more likely to produce conflict than clarity.

> **A prototype can have a bigger impact on the organization than on the problem.**

Are You Serious?

For precisely that reason, Jennifer Kemeny, an A. D. Little simulation designer who has done extensive systems-modeling work worldwide, swiftly focuses on the issue of commitment in her client engagements. "The question I always find myself asking new clients is, 'Are you serious?'" remarks Kemeny, who has facilitated hundreds of systems-dynamics sessions. "If you found a simulation credible, could you move on it?"

The perennial problem, she notes, is not that organizations lack the intelligence or capacity to learn from simulations and models. What they lack, she and her counterparts insist, is willingness to act on their learning. "What you often find is that organizations that say they need to change are really more interested in treating simulations as an intellectual exercise," says Kemeny. "They're doing it as a thing in itself, rather than as the beginning of an ongoing change process."

While there's nothing inherently wrong with intellectual exercises, says Kemeny, intellectual exercises are not how firms manage themselves. So before accepting a consulting engagement, Kemeny asks how the firm plans to use what they learn from the simulations. "If they say they're going to set up a committee," says Kemeny ruefully, "I know that I am going to have problems."

This question of seriousness directly relates to "Cui bono?" The seriousness of a firm's intentions often can't be gauged until it's clear who stands to suffer as the implications of a model become apparent. "What matters is whether they want to treat the learning as just another useful data point or as a rehearsal for change," Kemeny says. "If they're not serious about change, it doesn't matter how serious or valid the simulation is. [That's why] I pay much more attention to the culture of the client than to their technical skills." In other words, if a model is not going to be treated seriously, it is not going to be a serious model.

> **If a model is not going to be treated seriously, it is not going to be a serious model.**

What Do You Want to Talk About?

Instead of focusing on information creation, first-rate designers and engineers zero in on the human relationships their models affect. This is the crux of their value proposition.

Conversations with consumers

Former Chrysler vice chairman Robert Lutz emphasizes the value of prototyping for tapping into latent consumer appetites for design and marketing breakthroughs.

> *When we showed the early prototype for the new "big-rig-inspired" Dodge Ram pickup to consumer focus groups in the early '90s, the reaction was so polarized that the room practically vibrated with magnetism. A whopping 80 percent of the respondents disliked the bold new drop-fendered design. A lot even hated it! They wanted*

their pickups to keep on resembling the horizontal cornflake boxes they were used to, not to be striking or bold. According to traditional consumer research strategy, we should have thrown that design out on its ear, or at least toned it down to placate the hatemongers. But that would have been looking through the wrong end of the tele-scope, for the remaining 20 percent of the clinic participants were saying they were truly, madly, deeply in love with the design. And since the old Ram had only about 4 percent of the market at the time, we figured, What the hell! even if only half of those positive respondents actually buy, we'll more than double our share! The result? Our share of the pickup market shot up to 20 percent on the radical new design.

Lutz concludes with a key insight into the value of the prototyp-ing process: "What if we had gone ahead with a prototype that scored 7.5 [out of 10]? Examine the underlying data: If the 7.5 is a result of almost everyone scoring it a 7 or an 8 (with few or no 9s and 10s), it means you've got a guaranteed, inoffensive, uncontro-versial, bland, everybody's-second-choice loser! Nobody, in today's market, has to settle anymore for second choice."

Lutz and Chrysler wanted conversations and interactions that crackled with intensity. The goal of focus-testing the prototype was not to compile priority lists of favorite fea-tures and flaws but to polarize the passions of potential buyers. Emotional response mattered more than rational analysis. This is not to say that analytical conversations were unimportant nor that Chrysler ig-nored the critics. But Chrysler wanted to differentiate itself with designs that inspired passionate interactions. The company understood that it needed to manage its prototypes to create a different kind of conversation with its customers. Lutz credits this emphasis on prototyping for helping the twice-troubled auto giant to modernize its constrictingly centralized "product-planning" infrastructure.

> **The goal of focus-testing the prototype was to polarize passions.**

Conversations within the design team

Paul MacCready, the codesigner of the prize-winning human-powered *Gossamer Condor* aircraft, who is widely regarded as one of the finest aeronautical engineers of the post–World War II era, stresses that his quick-and-dirty prototyping philosophy is to spark design conversations that otherwise couldn't exist.

Stymied by the failure of computer-simulation models to explain the *Gossamer Condor*'s flight-control problems, MacCready decided to build a quick-and-dirty prototype:

> *I glued some balsa sheets to a stick to create a one-meter span "model" to resemble the* Gossamer Condor *geometry and pushed it through water in a swimming pool. The water, three orders of magnitude more dense than air, amplified the effects so much that my arm, serving as a poor man's wind-tunnel-force balance, was able to sense the rolling movement during yaw and the yawing movement during roll. This simple experiment suggested the answer that the computer could not articulate, and the final version of* Gossamer Condor *quickly emerged.*

"The swimming pool experiment broke the ice," says MacCready. "It gave [the design team] talking points that we didn't have before." The swimming-pool prototype sparked a conversation between MacCready and his collaborators that led to the final design breakthrough.

As MacCready himself comments, the purpose of the prototype was not to solve the control problem; it was to enable MacCready and his collaborators to rearticulate the problem. The design of the *Gossamer Condor* was a social process of relentless prototyping. "When you're inventing something new," says MacCready, "prototypes are a way of letting you think out loud. You want the right people to think out loud with you. The first prototypes usually don't answer your

"Prototypes are a way of thinking out loud. You want the right people to think out loud with you."

questions. But they are very good at starting the conversations that answer your questions."

Unprofessional conversations

Many managers misunderstand the conversations and interactions that even the most benign models can provoke. If the purposes of a business simulation are to "promote teamwork" and "build shared values," participants should not be surprised if the conversations turn personal. Teams consist of individuals with their own values, and values can't be shared until they're openly discussed.

Many organizations are unprepared for the nonprofessional and even unprofessional conversations their simulations provoke, Jennifer Kemeny reports. "It's like the old saying 'Be careful of what you wish for,'" she observes. "Sometimes the business issues can't be separated from people issues, and companies need to know how they want to deal with that." Rather than a "safe haven" for serious play, models and simulations can become an arena for intraorganizational conflict.

Managing models to promote cooperation or consensus or to evoke clash and creative tension is often only a secondary management consideration. But, as MacCready's and Lutz's experiences demonstrate, innovative prototypes invariably lead to innovative conversations. Innovative conversations can, in turn, enhance prototyping innovations. The management challenge is to create this kind of virtuous cycle of interactive innovation. That is always more difficult if management doesn't signal what kinds of conversations and interactions it wants to encourage.

Democratization of Design

Most commuters drive cars without ever looking under the hood. Managers use personal computers without knowing how to program. But is it enough for an organization to adopt models and simulations designed by experts? Or should the organization also help design and build them?

The emerging consensus is that the act of designing models or simulations is essential to understanding their use. The most valuable managerial insights arise from both designing the rules and playing the game. "I learned the most from designing the simulations," says Clark Abt, "and the least from running the damn things. But you have to do both. . . . Instead of delivering simulations to clients, we should have gotten them much more involved in their design. But that was very difficult to do in the 1960s and 1970s. The technology wasn't really there, and frankly our biggest clients in government weren't all that interested in collaboration. . . . We thought of ourselves as the experts in simulation design. That's why our clients selected us."

David Lane, a British management professor who has consulted widely in simulation modeling, argues that organizations must be full participants in creating the models they use. Experts shouldn't create models and simulations; they should facilitate collaborative creation. "Rather than attempting to take the position 'I am an expert in techniques that will teach you about your business,'" says Lane, "the consultant should offer a process in which the ideas of the team are brought out and examined in a clear and logical way. . . . *Clients' ideas must not just be in a model, they must be seen to be in a model* [emphasis in original]."

Engineering organizations have found that nonengineering managers and marketers want to play with CAD software to test their own product ideas and enhancements. Such "amateur CAD" signifies a growing democratization of design promoted by pervasive and accessible modeling technology. The changing nature of the modeling medium is forcing design professionals to manage the prototyping efforts of design amateurs. Experts in engineering, finance, manufacturing, logistics, marketing, operations, and other value-adding processes of the firm are

"Amateur CAD" signifies a growing democratization of design.

being called on to shift from building better models for the firm toward building better models with the firm. The declining cost and rising importance of prototyping is broadening the community of designers.

Playground or Rehearsal?

Thomas Schelling, an economist and political scientist who for decades ran classified nuclear-war-simulation games for government officials, once observed how hard it was to get his combatants to "go nuclear," even with the most outrageous provocations. "People weren't playful or adventurous when it came to using nuclear weapons," Schelling recalls. "They took their [simulated] roles and responsibilities very seriously. People wanted their side to win but I can't say they treated this like a game."

Of course, there is no more apocalyptic scenario than a nuclear holocaust. But Schelling's experience highlights a key issue for modeling: Will an organization learn more from treating a model or simulation as a slice of reality or as a game? Is it more advantageous to try to reproduce reality or to try to create it? Treating a model as a rehearsal medium calls for a different level of commitment than using it to test the limits of the imagination. In the former, respecting the rules may be paramount; in the latter, the rules were made to be smashed beyond recognition. Sometimes the two are indistinguishable.

Will an organization learn more treating a model as a slice of reality or as a game?

Arie de Geus has described how this modeling conflict manifested itself at Royal Dutch/Shell:

> *More often than not, when we began to put two or three managers'*
> *explanations together into the same model, calibrated against one*
> *another, we saw a [reversion] to the old, suspicious attitudes about*
> *computer "black boxes". . . . It would be impossible to get the group*

> *of managers to start* playing. . . . *They became critics of the model.*
> *They passed hours querying its assumptions, pointing out omissions,*
> *criticizing the modeling techniques—anything but learning about*
> *their environment! . . . Managers wanted the model to resemble not*
> *just reality, but their own assumptions about the external reality. If*
> *in doubt, they would refuse to play.*

De Geus wanted his managers to play with the model on its own terms; the managers refused. Classically trained musicians have difficulty playing jazz. Improvisation within a set of themes requires a fundamentally different relationship to the music and to one's fellow musicians than consistently interpreting a score. As Carnegie Mellon design engineer Dan Droz has observed, "The idea that you can 'play' your way to a new product is anathema to managers educated to believe that predictability and control are essential to new-product development." Often, however, playing with prototypes and simulations is precisely how breakthrough innovations occur. The problem is figuring out when it's most productive to improvise and when to rigorously follow the score.

Every organization that uses models, simulations, and prototypes has to choose how playful or serious it wants its interactions to be. Certainly, airplane-cockpit simulations for pilots are treated with the gravest seriousness. But even here, safety researchers test pilots' abilities to improvise under stress; confirming that pilots know how to respond to predictable emergencies is not enough. Similarly, the National Research Council's crisis-management simulations recognize that all contingency plans require improvisational flexibility.

The question boils down to, What should the model ask of participants? To think or to act? To improvise or to follow? To become introspective or communicative? Does it ask for answers or for questions? To be yourself or to look at the world through different eyes?

MIT's Lincoln Bloomfield, who like Schelling ran hundreds of "political exercises" for the State Department and other federal agencies, says that getting officials to role-play was one of the most valuable aspects of the simulations. "It was always interesting to see how people behaved in the roles of their enemies," says Bloomfield.

"Most people did a very responsible job of role playing." Unlike Schelling, Bloomfield did observe gamelike improvisations designed to challenge policy assumptions. "The interactions, the surfacing of plausible alternatives, mirrored the decision-making process of policymakers at their best," says Bloomfield. One of his most surprising observations is that "scenarios prepared by experts are most often fundamentally incorrect." The simulations often did a better job of generating plausible geopolitical policy options than the various State Department and National Security desks, he asserts. The line between "playing" and "rehearsing" policy was vanishingly thin.

> **"We don't believe in role-playing . . . people have enough trouble being themselves."**

By stark contrast, Booz-Allen partner Mark Frost, who runs business-simulation games for the giant consulting firm, flatly states, "We don't believe in role-playing . . . people have enough trouble being themselves." Frost argues that most managers need simulations to test their own perspectives and beliefs, not to pretend to appreciate a competitor's.

Creating Narratives

A prototype isn't merely a prop on the organizational stage; it is a character in a marketplace narrative. A computer simulation isn't just a bundle of analytical software; it becomes an animated film with protagonists, conflicts, climaxes, and denouements. As European design critic John Thackara brilliantly observes, postindustrial innovation isn't about the design of objects; it's about the design of experiences.

> **Postindustrial innovation isn't about the design of objects; it's about the design of experiences.**

The best designers and innovators have shifted their focus accordingly, from prototyping products and services to prototyping experiences and interactions. They're using prototypes to tell stories, and they're telling stories to develop prototypes. "We treat our prototypes like they're characters in a big-budget movie," says industrial designer John Rheinfrank, who goes on:

> *So we use sketches and storyboards and videos and every representa-*
> *tional medium possible to animate what we're designing. . . .*
> *We want to see what roles it can play. . . . It used to be we would*
> *improve the prototype by iteration; now we improve it by telling new*
> *stories and iterating the prototype around the versions of the*
> *stories. . . . The scenarios become the common ground for our design*
> *discussions. . . . One of our big questions is deciding who gets to tell*
> *the stories of the product. . . . At Xerox, one of the best stories we*
> *heard came from the service people, and that had a big impact on*
> *the design.*

IDEO's David Kelley agrees: "We care as much about great interaction design as we care about great mechanical engineering. The problem is, calculating statics and dynamics is easier than designing interactions."

Design-driven firms like IDEO, Walt Disney, Chrysler, and Gillette all use stories and scenarios to turn the prototyping process into performance art. IDEO's engineers prepare product storyboards that could pass muster at Disney. Gillette's industrial designers have prototyped razors, toothbrushes, and coffeemakers based on ethnographic research that turns household life into anthologies of product-based narratives. These companies are no longer merely prototyping innovative products; they are prototyping innovative stories about *interacting* with products.

IDEO's California design studios resemble an animation studio. Inspired by Disney animators, Rheinfrank goes into client companies and turns rooms into workspaces where product storyboards hang on the walls and rapid prototypes lie scattered on tables. "Design spaces are where we take turns telling stories, making storyboards, and building prototypes," says Rheinfrank. "Sometimes we take a prototype and create a storyboard around it; other times, we do a sketch that gets us to building another model. All our discussions are facilitated by either the prototypes or the sketches or the storyboards."

But as Rheinfrank and Kelley have observed, engineers and innovators aren't always the best storytellers. The ability to tell a tale around a prototype is often hard-won. As Keith Johnstone, a teacher of theatrical improvisation, remarks, "If I say, 'Make up a story,' then most people are paralyzed. If I say, 'Describe a routine and then interrupt it,' people see no problem." In fact, interrupting routines, which Johnstone sees as the essence of improvisational story creation, could just as easily be defined as the seed of innovation. Interrupting a simulation, or causing a slight alteration of a prototype to interrupt the routine of its planned use, is a powerful technique for generating new narratives.

> Interrupting routines could be seen as the seed of innovation.

PricewaterhouseCoopers partner Win Farrell agrees that simulations are "a form of theater," but he prefers another technique to spur improvisational narratives: conflict. "Framing a dichotomy intensifies interaction," he observes. Farrell, who uses sophisticated mathematical modeling techniques to devise simulations for clients in telecommunications, entertainment, and retailing, stresses that clients get more out of models if they are challenged to come up with multiple stories for them. For a telecommunications simulation, he typically splits a management group into several teams and asks each to run the simulation with explicit strategic dichotomies in mind: maximizing profit versus maximizing market share, or building infrastructure versus reselling capacity, or organic growth versus acquisitions growth. "When people overhear their rivals at the other tables discuss their strategic trade-offs, they start to get competitive," says Farrell. "We've built a market in strategic conversations around the simulation." The result, he says, is a set of stories about which trade-offs within each dichotomy make the most sense. "The search for 'the best trade-off' is an illusion," says Farrell.

"Simulations aren't about finding optimal solutions; they're about discovering what stories make sense."

Consultant Jennifer Kemeny notes that, in her simulation-facilitation work, she tries to move clients away from discussing, "What's best?" toward, "What's most robust?" Firms frequently use models and simulations to try to determine optimal solutions to problem, she notes, but optimal solutions are often unsustainable. On a risk-adjusted basis, best solutions are often more expensive than robust solutions.

> Sometimes "What's best?" is a less useful question than "What's most robust?"

For example, an optimal simulation solution might call for a logistics-delivery system managed by a hub-and-spoke typology, like that of American Airlines. A more robust solution might be to better manage point-to-point direct deliveries, in the manner of Southwest Airlines. Hub-and-spoke may technically be the optimal solution, but the cost of occasional hub disruptions could wipe out the presumed overall benefits. In other words, the "What's best?" story has different heroes and villains from the "What's most robust?" story.

These stories about stories bring to mind Kierkegaard's observation that "the sign of a good book is when the book reads you." In the virtual environments of modeling media and prototype-driven conversations, "The sign of a good simulation is when the simulation plays you."

Hot Washup

"Hot washup" is military slang for reviews of a military war game. Why did the participants make the moves they did? What did the good moves have in common? What explains the failures? In the armed forces, war-game exercises and simulations are assiduously picked apart and analyzed. Command behavior and command decisions are rigorously reviewed. The results can mean the difference between a promotion and a career plateau. "The military is pretty good at learning from its war games," says James Dunnigan, who has designed military simulation games since the 1970s. "There's still a lot of politics, of course, but this is one of those areas where the private sector could pick up a few pointers."

In medicine, there is a strong tradition in teaching hospitals of "post-mortems" and case reviews. Notwithstanding a bias toward examining critical failure factors rather than critical success factors, the best hospitals attempt to create an environment characterized by well-documented decisions and ongoing review. Cases are reviewed with the expectation that learning will take place. This rigor is scarcer in business. Management surveys reveal that surprisingly few firms formally compile lessons learned from their prototyping experiences. Lincoln Bloomfield and Thomas Schelling both deplore the national-security establishment's laxity about formally capturing the lessons of the political exercises they facilitated. Too often, says Bloomfield, the simulation itself is treated as the goal. Its implications are managed as an afterthought.

> **Too often the simulation itself is treated as the goal, and its implications as an afterthought.**

Formal review processes do have a few notable business champions. MIT Professor Michael Cusumano's research on Microsoft affirms that the software giant rigorously employs reviews to plan new versions or upgrades of software packages. Similarly, GE Capital—long the profits powerhouse of Jack Welch's General Electric conglomerate—constantly revisits the financial models used to justify its investments. "God help you if you cannot document every reason why you chose to modify the model," says one GE Capital executive. "This is a culture that makes people totally accountable for how they use [models]. We are expected to share what we've learned from our models and we do. . . . You could never have gotten away at [GE Capital] with what they got away with at Kidder [GE's failed investment-banking subsidiary]."

But just how introspective is introspective? Most organizations wouldn't hesitate to videotape a customer focus group interacting around a new product prototype. But how many design teams videotape themselves interacting around their proposed innovation? Why don't more organizations use videotape and other recording media to monitor their own design interactions? Could individuals learn to improve their own performance by watching their behavior around prototypes? These questions will become moot as firms recognize that more creative innovation requires more creative introspection. CAD systems make it ever easier to track the design evolution of product innovations. The ability to peel back and audit what design decisions were made, and when, has become a critical part of the design introspection/improvement process. "We look back at our CAD designs all the time," says IDEO's David Kelley. "Sometimes we do it with our clients; sometimes we do it by ourselves." This retrievable and auditable record of the design process, says Kelley, has become a core aspect of the firm's organizational memory.

> Could individuals learn to improve their own performance by watching their behavior around prototypes?

But the ability to audit the evolution of a prototype is not enough. Just as world-class athletes review their performance in practice, world-class managers and innovators will need to review their performance in managing design and innovation processes. Organizations won't just learn from their most creative interactions; they will also learn from recording and reworking their most creative interactions.

MEASURING PROTOTYPING PAYBACKS

When you can measure what you are speaking
about, and express it in numbers, you know
something about it; but when you cannot measure
it, when you cannot express it in numbers, your
knowledge is of a meager and unsatisfactory kind.
Lord Kelvin

Consider a high-profile but niche global industry now utterly dependent on rapid prototyping and simulation technologies. Capital-intensive, nationalistic, and ruthlessly competitive, this high-tech marketplace features feisty entrepreneurs challenging well-heeled corporations. Government subsidies and tax breaks mysteriously materialize then disappear. Top design and management talent hop, skip, and jump to the highest bidders. The world's leading universities provide state-of-the-art supercomputing and software services to test design concepts.

Technology has become the bit-stained battleground where unfair advantages are hunted down and captured for further exploitation. Where experienced intuitions once ruled supreme, empirical experimentation and testing now dominate. Today, no strategic decision goes unmodeled. No competitive contingency goes unsimulated. Yes, charismatic leadership and human capital matter more than ever. But success requires leadership in innovative technology, not just leadership of innovative people.

So the key players delicately balance their newfound tradition of digital innovation with opportunistic episodes of industrial espionage. Highly regulated by a multinational overseer, this industry is as creatively litigious as it is creatively innovative. And why not? The stakes are very high. The marketplace is unforgiving: winner takes all.

No, this is not some bizarre corner of the biotech industry or newly deregulated slice of telecommunications. This is the real world of the America's Cup sailing competition. An America's Cup syndicate may spend more developing a high-tech yacht than a Silicon Valley venture-capital firm invests in its entire portfolio of start-ups. The difference is that the boat is far more rigorously designed than any of the start-ups. Funded by wealthy yachtsmen and even wealthier corporations, the America's Cup has become the world's most innovation-driven sports competition. Passion still matters more than profit, but winning matters most of all.

Despite the unusual profile of its market forces, the America's Cup is an astonishingly relevant microcosm of the mission-critical issues surrounding prototyping and simulation. More surprising is that the savviest America's Cup sailors seem far more innovative and attentive to performance metrics than their landlocked counterparts. In fact, the most successful America's Cup challenger and winner of the 1990s brilliantly leveraged itself around its key prototyping metric.

The 1995 America's Cup is a case study in prototyping-metrics methodology.

The 1995 America's Cup competition is a compelling case study in the strategic importance of prototyping-metrics methodology. That year, Team New Zealand's *Black Magic* yacht captured the America's Cup with only one loss. This unexpected triumph by an underdog was the direct product of Sir Peter Blake's determination to build a team culture around experimentation and simulation. The single most important strategic decision Sir Peter made was to insist that the syndicate fund construction of two boats. These boats were to be—with some marginal differences—near-identical twins. The reason was straightforward: Team New Zealand (TNZ) wanted to measure the impact of every design change implemented on the water. This two-boat strategy was not unique to New Zealand, but no other country had ever invested in a prototyping-and-simulation infrastructure as thoughtfully and extensively as Sir Peter's *Black Magic* syndicate.

This approach, though far more expensive in hard-dollar terms than a single-boat strategy, would be a clear and unambiguous way to measure "return-on-innovation" investments. There would be a metric that mattered and that could be managed. In effect, TNZ wasn't just going to build faster boats: it was going to build an internal marketplace in which people and ideas could be tested on the rough seas of competition.

Doug Peterson, a member of the winning 1992 America's Cup team, was brought on as TNZ's lead designer. The syndicate also recruited simulation-software experts to create virtual environments for Peterson's designs. Peterson's concepts initially defined the design, but the focus radically shifted after the yachts were built. Thousands of simulations were run to evaluate the impact of possible design iterations. These simulations were not run remotely on some supercomputer hundreds of miles away, but at a workstation cluster not fifty yards from the dock.

The physical boats complemented the virtual boats. Every day, one of the two physical boats was modified to test a possible design change. The two boats then raced to see what impact, if any, the change had made. "We could not have improved the speed as we did without racing the two boats," Blake insists. "There was no honest way of testing the changes with only one boat." The crew could also see the impact of the changes, and they were encouraged to make suggestions. Their ideas would first be tested virtually, and the best would swiftly find their way onto the boats for testing. On a good-weather day, over a half-dozen changes could be effectively tested.

One could argue that there were at least three *Black Magics* sailing in San Diego: the two yachts on the water and the virtual yacht sailing in the Sun Microsystems computers. The defending U.S. yacht, by contrast, had been designed by a state-of-the-art supercomputer capable of simulating thousands of redesigns. But this machine was located hundreds of miles away, and its interface was not configured for interactive iterations; time delays were significant. Furthermore, the U.S. team had only one boat. Thus there was no way the U.S sailors could quickly and reliably assess proposed design modifications.

The result was a sweep: the American boat was utterly outclassed. Paul Cayard, its skipper in the final race, commented, "I've been in some uphill battles in my life. But I've never been in a race where I felt I had so little control over the outcome. It's the largest discrepancy in boat speed I've ever seen."

Mean-Time-to-Payback

Sir Peter's successful pursuit of his two-boat strategy within a rapid-response prototyping infrastructure transformed the America's Cup competition. The ingenuity exceeded the investment. Other competitors spent far more, but they never truly measured their returns on design innovation. An honest metrics marketplace made the difference: the New Zealanders created a healthy prototyping culture that swiftly and accurately measured the impact of its design innovations.

Team New Zealand tested more design iterations faster, but that was only half the battle. The fundamental breakthrough was measuring the performance paybacks of those iterations almost as fast as weather conditions permitted. The value wasn't just in rapid prototyping; it also resided in TNZ's ability to test and assess the performance results of prototyping rapidly. The key to managing prototyping metrics is to understand that prototyping and simulation are means, not ends. The role of prototyping metrics isn't just to improve the prototype but to create a winning product or service.

> **Other competitors spent far more, but they never truly measured their returns on design innovation.**

Consequently, the mantra that speeding innovations to market ASAP is the key to competitiveness is dangerously misleading. High-technology companies and innovators who redesign themselves around that belief may be rapidly accelerating down the wrong side of the innovation speedway. The link between rapid prototyping and speed-to-market may be precisely the wrong innovation ideal to champion.

Speed-to-market might well be the most misunderstood management metric around. It's true that companies from Boeing and Chrysler to Intel and Cisco reap competitive advantage by slashing months from product-development cycles. But that's unlikely to be the real reason they're leaders in rapid innovation. The innovation metric that matters most in a hypercompetitive and volatile marketplace measures a different kind of speed; its focus is the needs of customers and clients rather than the speed of the innovator. The single variable that customers and clients care most about is their own mean-time-to-payback—the speed at which they believe they will get payback from purchasing a product or service.

The customer's perceived mean-time-to-payback, not the innovator's speed-to-market, effectively determines which innovations will dominate their markets. The faster the payback, the better the chances that an innovation can create a meaningful market for itself. No other metric is more important. Indeed, the business history of the digital innovation is a story of ever-accelerating mean-times-to-payback for the people who purchase silicon and software. Dan Bricklin, who cocreated VisiCalc, attributes the success of the spreadsheet that launched the personal-computer revolution to the fact that financial analysts believed that the system paid for itself in under a week. The number crunchers so hated doing their spreadsheets by hand that—even when they factored in the price of a new Apple II on which to run VisiCalc and the effort required to learn to use both—the rapid mean-time-to-payback virtually guaranteed the sale. Lotus Development's Mitch Kapor paid close attention to that lesson. So did Bill Gates. So did James Barksdale when Netscape created the browser market. Intel's microprocessor business model, pioneered by Bob Noyce and aggressively extended by

Andy Grove, also incorporates the principle of swift mean-time-to-payback.

Payback as a prism

The Polaroid camera, Federal Express overnight delivery, Xerox's photocopier, cellular phones, Intuit's Quicken, and even McDonald's hamburgers are all obvious examples of innovations with breathtakingly swift mean-times-to-payback. All these products and services make it easy for customers to calculate just how quickly they will pay for themselves. Customers don't just want to buy features and benefits; they want to buy immediate payback. Their purchasing priorities embody actress/novelist Carrie Fisher's sage observation that "instant gratification isn't soon enough."

Viewing innovation through the prism of payback is a transforming experience. All kinds of business behavior and market opportunities become more transparent. What's the easiest, most obvious, and laziest way to reduce mean-time-to-payback? Cut prices. Why are ease of use and quality interface design important? Because products and services provide faster payback if they're easier to use. Consider the rise of leasing in the multibillion-dollar U.S. automobile marketplace. Look at the financing and training schemes and customer-support services increasingly bundled into innovative products. Companies aren't positioning and packaging their innovations just to make them more valuable, but also to make them more valuable even faster.

> **Customers don't just want features; they want immediate payback.**

The mean-time-to-payback perspective also helps explain the idea of increasing returns due to "network effects" championed by economists like Brian Arthur of the Santa Fe Institute. The millionth telephone or television owner will likely have a far faster mean-time-to-payback than the thousandth. Consider the explosive rise of the Internet: the mean-time-to-payback for the next browser—

or the next server—accelerates as more people jack into the Net. This concept goes a long way in explaining the Internet strategies of Yahoo!, eBay, and Microsoft. For a growing segment of the global population, the Internet's mean-time-to-payback is almost instantaneous.

Customers' payback metrics

Customers also have individual metrics for payback that go beyond price. When Computer Associates, a multibillion-dollar vendor of enterprise software, tried to give away copies of its personal-finance software to crack the consumer-software marketplace, it barely nicked Intuit's domination of this category. Why? Because if it takes three or four hours to figure out how to use the software, customers can't count on a swift mean-time-to-payback. As recording devices, VCRs didn't offer swift mean-time-to-payback—we've all seen those blinking 12:00s. Their rapid payback stems from the rise of Blockbuster and other video-rental places that make it cheap and easy to rent movies.

Payback isn't just a matter of money and time; fundamentally, it involves behavior change. That's why a comparatively simple computer language like Visual Basic is far more popular than C++. That's why complex CAD programs like CATIA are worth their premium price to global giants like Chrysler and Boeing: cutting mean-time-to-payback on a multibillion-dollar airplane or automobile development offers a big win.

Innovators need to become highly sensitive to their customers' payback calculations. Then they must design their products and services accordingly. The dominant innovation issue isn't, "How do I make my innovation better or more valuable or bring it to market faster?" It's, "How do we cut our customers' mean-time-to-payback?" Clearly, one's customers—not one's prototyping processes—should determine the appropriate rate

> **The dominant issue is "How do we cut our customers' mean-time-to-payback?"**

of innovation. What use is a speedy rate of clever innovation that outstrips the customer's desire or ability to assimilate it?

Outside-in metrics

The most effective prototyping and simulation metrics work from the outside in rather than the inside out. That was certainly true for *Black Magic*. Prototyping metrics that reflect customer experience are more likely to be tightly coupled to value creation than metrics focused on the prototype itself.

Prototypes constructed around the customer's perceptions of value—price, ease of use, scalability, reliability, versatility, and the like—offer a different approach to measurement than those built around the organization's perception of value. After all, innovative products and services don't necessarily fail because they are poorly designed or engineered—they fail because customers and clients don't see fit to pay a premium for them. Models and simulations enable organizations to swiftly test the value propositions of innovations they think might matter.

Innovative products fail because customers don't see fit to pay a premium for them.

When aligned with appropriate metrics, models and simulations offer the most powerful possible way to discover and deliver unique value. Whatever the goal, making models measurably accountable dramatically improves the chances for marketplace success.

Mean-Time-to-Implementation

But which metrics? What paybacks? Measurement for its own sake is as managerially wasteful as prototyping for the sake of prototyping. For Royal Dutch/Shell, a company that has made a strong operational commitment to prototyping, these questions were a constant torment. The oil giant's culture was predicated on a belief that the benefits of prototyping could be credibly measured. That learning for its own sake was justification enough or that professional development required top management to engage in sophisticated make-believe was explicitly rejected. Shell was a business, not a university.

Neither fools nor charlatans, Shell's modeling mavens didn't claim that top managers who played with simulations would become better decision makers or smarter leaders; they focused instead on behavior. "Because one cannot measure the quality of decisions, we decided we would measure speed: How long did it take to get from the *perception* of a changed external reality to the *implementation* of a fundamental shift of operation?" reports Shell's Arie de Geus. "When we measured, we tended to find that the process of learning had been accelerated by a factor of two or three; it was now two or three times as fast to implement a new internal system"—in other words, mean-time-to-implementation, a useful variant of the mean-time-to-payback metric.

De Geus cites the company's rapid adoption of an oil-trading operation as a tangible illustration. "The vertical integration that had held sway for 40 years, under pressure from nationalization in the Arab world, was disintegrating fast," he writes in *The Living Company*. "There had been spot markets in oil trading before, but oil companies did not immediately recognize their need to shift their managerial approach: from optimizing the flow of oil within the company only to being willing to say, 'Every drop of oil that I have is in principle for sale—not just to our own company but to anyone.'"

Shell's top managers had, however, been simulating an oil-commodities trading floor using the STELLA computer language. Interactions around iterations of this "virtual-pipeline" simulation had made the company acutely aware that a new marketplace would require new behaviors and a new vocabulary of value management. In response to changes in the external marketplace, and the fact that rivals like British Petroleum already had their own trading floors, Shell moved to create a trading floor. "It took us only six to seven months," de Geus recalls. "Within the following year or two, the amount of oil traded in that system rose to become 40 to 50 times as much oil as actually, physically, moved through the real system of refineries and tankers. In the past, similar decisions had taken 18 months or more."

For Shell, simulation was as much a medium for accelerating implementation as it was a learning tool. The metric was speed. Whether the STELLA simulation saved Shell six months or twelve months is an interesting question, but the key question is, What were the legitimate, measurable alternatives? What other forms of preparation could have comparably prepared the company for this new marketplace? Did they offer comparable metrics to measure potential success? Shell does not claim that building better prototypes consistently builds better managers. Its claim is that better interactions around prototypes make for faster and more sophisticated implementations of the decisions that managers make. In other words, prototypes can both streamline and refine managerial interaction. Shell's payback metric offers a discipline and a test for any organization eager to explore a potential reality.

> **Simulation was as much a medium for accelerating implementation as it was a learning tool.**

How to Slow Down a Browser

Sometimes speed can be the enemy of payback. When Steelcase, the Grand Rapids–based multibillion-dollar office-furniture manufacturer, decided to market its "office systems" differently, the company redesigned its staid sales showrooms. Working with IDEO, Steelcase modeled interactions between potential customers, salespeople, and the office products themselves. Drawing heavily on ethnographic research, showroom observations, and interviews, the IDEO/Steelcase team developed concepts that would turn the proposed New York facility from a "pristine, static showroom into an active resource," in the words of Aura Oslapas, who ran IDEO's environments practice. "We . . . presented interactive scenarios that described initial concepts for a customer journey through the building," Oslapas recalled. "We identified a broad continuum of visitors, from those focused on physical attributes and product design to those focused on less tangible issues. . . . . We created an 'experience map' to sort and link all the possible interactions. . . . We developed three main areas of interaction—settings, pods, and sandboxes—to allow sales staff and customers to access information in creative and dynamic ways."

Briefly, *settings* are work environments, *sandboxes* are places for play (for instance, the Materials Sandbox displays finish materials and colors), and *pods* present alternative office systems. Scale-model trolleys allow customers to visualize possible layouts and configurations in their own office settings.

Steelcase designed the settings, sandboxes, and pods as prototypes; they evolved through interaction. And they were designed in turn to encourage potential Steelcase customers to prototype a Steelcase office system. In other words, Steelcase built a prototype environment that enabled its customers to build their own prototype environments.

What metric best captures the impact of such a radical redesign? Obviously, sales and traffic are important indicators. But when you are fundamentally redesigning your sales environment to optimize inter-actions, as opposed to merely presenting and exposing the products to potential buyers, time is of the essence—though not in the usual way. Typical Steelcase showrooms had barely held the attention of customers for forty minutes. This became an important benchmark. In the course of redesigning the New York facility, the challenges became both capturing and holding the customer. The marketing axiom was that the longer customers stayed, the more likely they would be to make purchases. How do you create an environment that entices people to stay longer? IDEO's storyboards and

> **Steelcase built a prototype environment for customers to build their own prototype environments.**

prototypes for the New York facility ultimately focused on these kinds of questions. The "experience map" was about transforming the customer experience from passive to interactive.

The result? Floor traffic and corresponding sales exceeded expectations. The New York facility was a success. But the metric that particularly mattered to Oslapas and her design team was that potential customers now spent roughly two hours in the facility. Customers' behavior had changed. Steelcase had transformed a showroom into a time machine with a twist: the provocative proto-typing payback was getting people to slow down.

Markets in Metrics

Weight as a currency

In the business of aviation, weight is money. When Boeing negotiated the performance requirements of its 777 jet with United Airlines and British Air, the most unforgiving design specification was the aircraft's weight. According to Boeing engineer Henry Shomber, the airlines demanded performance contracts that would financially punish Boeing for excess fuel costs over the life of the 777 due to excess weight. The airlines weren't just going to pay for performance; they wanted to pay for economic productivity. The 777 was designed to be a product of state-of-the-art economic engi-neering as well as superior technological engineering.

The 777 represented the most dramatic effort in Boeing's history to abandon Mylar in favor of digital design media. Using these new technologies, Boeing could build a new jumbo airliner better, faster, and cheaper than any other airframe it had ever built. But if the plane didn't hit the desired weight targets, it could prove unprofitable. Weight is money.

Boeing's management recognized that a control-system param-eter—a prototyping metric—would be necessary to assess quickly the economic impact of design decisions that would change the plane's weight. Interactions around iterations of the 777's prototypes would have to address the challenge of the aircraft's avoirdupois.

The classic command-and-control approach would be to appoint a "weight czar" or assign the program manager to make all weight-related decisions. The obvious danger of such an approach is that schedules slip as the number of decisions increases. Worse, the weight czar or program manager may not be fully versed in the technical trade-offs associated with letting some subsystems gain weight while keeping others on a strict diet. The subtle trade-offs between local and global optima may not be readily grasped by a single mind.

A more sophisticated approach, more typical in the aerospace industry, is to establish a "weight budget." Just like cost budgets for subsystem developments, which are negotiated, tracked, and audited, there are weight budgets that managers are measured against. But Boeing didn't just create a weight budget; it created a cleverly regulated "weight economy," with its own carefully calibrated pricing system and explicit incentives for designers and engineers to, quite literally, weigh the consequences of their actions.

The company determined by applying its own financial rules the economic costs of missing the overall weight objective. Boeing then devised a set of decision rules, which it rigorously communicated right down to the level of the individual designer. Depending on the subsystem, an individual designer could make a weight-reducing design decision that would cost up to $300 per pound per plane without approval, and up to $600 per pound with a supervisor's approval. With the signature of the program manager, a designer could spend up to $2,500 per pound. Boeing's prototyping metric effectively put a price tag on weight reductions. This approach has clear

Boeing's prototyping metric put a price tag on weight reductions.

advantages over a traditional weight budget. That approach would offer no incentive for designers of a subsystem that was within its weight budget to work any harder or more creatively. But if your system had gotten obese, you might resort to costly choices and trade-offs. This might mean, for example, that the hydraulics team buys weight for $2,500 a pound while the engine team neglects straightforward opportunities costing only $300 per pound because they are still within their budget.

"[These] decision rules are powerful," asserts Donald Reinertsen, author of *Managing the Design Factory* and an expert on speed-to-market innovation, "because they optimize at the system level, . . . avoiding the problems of locally optimal but globally suboptimal decisions." In other words, the creation of a 777 "market economy" in design simulation and prototyping led to a more sophisticated set of design trade-offs. Ultimately, the physical 777 didn't end up unprofitably overweight. Boeing's internal market forces exerted a powerful and necessary design discipline. The world's most sophisticated CAD technology allowed for all manner of design trade-offs to be considered; Boeing's prototyping metric—the economic value per system pound—better defined the value context of those trade-offs.

Microeconomies

The message is clear: the most effective way to reap added value from prototypes is to create markets in useful metrics. Whether the currency is head-to-head competition, time, money or weight, prototypes can create a marketplace in metrics. Design trade-offs can be bought, bartered, and traded. So can implementation trade-offs.

Design trade-offs can be bought, bartered, and traded. So can implementation trade-offs.

Microeconomic techniques will increasingly be brought to bear on the management of firms' prototyping-and-simulation economies. To wit, do the benefits of increasing the number of iterations per unit of time outweigh the costs? What metrics determine the answer? Should different "prices" be attached to modifying a prototype at different stages in the development cycle? On large corporate software projects, it's not unusual for a "change control board"—technical experts and business managers—to try to regulate the introduction of new features, prune troubled functionality, and "tax" potentially expensive modifications. At what point does such regulation stifle innovation? At what point does deregulation degenerate into prototyping anarchy?

These are not fanciful questions. A recent review of the management literature in the authoritative *Administration Science Quarterly* questioned the effectiveness of simulations in business because of the absence of credible and replicable metrics:

> *The same questions about effectiveness apply to the management games or simulations for improving managerial or organizational performance that have become increasingly popular during the past two decades. . . . Very few of these games and simulations have been subjected to sound evaluation research. . . . [There is] no published evidence that leaders who perform well in management games and simulations will improve their leadership performance in their real-life jobs. Our own studies did not even produce any significant or substantial correlations between a group's performance in one simulation and the same group's performance in a similar simulation.*

The plural of anecdote is not data. Too often, even the most prototyping- and simulation-intensive cultures don't use metrics to measure their modeling practices. That is a mistake. Metrics are essential for successfully managing change in modeling culture and practice.

The plural of anecdote is not data.

As digital technologies become the dominant prototyping media, it should become ever easier and cheaper to track key indicators. This is a central message of Chapter 3 on modeling culture. There are benchmarks and measures that every organization can use to interrogate itself about which practices offer the greatest return on resources invested. These metrics hold the key to the strategic introspection necessary to manage an innovation culture.

All organizations need to know what their prototyping budgets are. Every enterprise needs to have some awareness of its mean-time-to-prototype. Just how many prototypes are iterated per unit of time? What is the typical number of modifications per prototype per unit of time? What are the costs of those modifications? What is mean-time-to-demo internally? Externally?

These questions, like the examples of successful prototyping metrics presented earlier in this chapter, are but the beginnings of wisdom. Different prototyping cultures will spawn different metrics and marketplaces, just as different enterprises rely on different accounting methodologies and risk-management practices. But it's a sure bet that these differences will help shape competitive advantage and disadvantage as prototypes migrate from the internal economy into the external marketplace.

Software guru Tom DeMarco makes three telling and transcendent observations about software metrics:

- Measure benefit or don't measure anything.
- Measure for discovery.
- Be suspicious of any finding that confirms your darkest suspicion.

DeMarco's three admonitions point the way toward aligning organizational culture with the relentless demands of marketplace performance. They are appropriate metrics for selecting appropriate metrics.

9

G O I N G   M E T A

Evolution as a
Business Practice

The reason why we are on a higher imaginative
level is not because we have finer imaginations, but
because we have better instruments.
Alfred North Whitehead, 1925

The great thing about the Web: You can bring up
a product up on your site—if people hit on it,
you call it a product; if they don't, you call it
market research.
Jim Barksdale, former CEO of Netscape

Internal versus External Model Markets

Innovating for market share requires a different investment portfolio than innovating for profit. Growing margins demands different innovation investments than growing revenues. Managers seeking to anticipate market demand treat prototypes differently than managers who prefer to be market-responsive. Fast followers can be even more innovative than first movers, but their internal marketplaces will not be comparable. There is a world of difference between firms that create innovations from scratch and those that treat competitors' products and services as prototypes to be reverse-engineered. Being first-to-market no longer assures either a market premium or a sustainable competitive advantage in the face of fast followers whose innovation infrastructures are geared to reverse engineering. Fast followers are not the same thing as free-riders; fast followers often come up with enhancements that represent advances over the original offering. But as modeling media become more commoditized, firms of both types will redesign their incentive and innovation infrastructures to seek new value-creation venues. They will realign their internal modeling marketplaces to better reflect the value they want to offer their customers. It seems certain that the innovation enterprise will increasingly draw upon microeconomic and financial tools to manage its "prototyping economies." We will be using models to manage our models. "Options pricing" and the "capital asset pricing model" financial engineering techniques will be imported to the innovation infrastructure.

For example, an innovator that believes customers most value customized choice at reasonable cost will manage its prototypes to optimize most-choice/best-price options; customers will select from a broad menu of options to customize their product. Another innovator that believes its customers value the ability to codesign will invite active customer collaboration on the proposed product. Both firms offer genuinely customized products. But ordering a meal is not the same thing as being invited into the kitchen. The value propositions and business models are fundamentally different, and the prototypes themselves reflect fundamentally different design sensibilities even if the ultimate products are quite similar.

By contrast, consider a significant digital-design innovation trend. Carliss Baldwin and Kim Clark of Harvard Business School argue that modular design of products and services is becoming commonplace. Modularity means that component and subsystems manufacturers can fit modules together into fully functioning products by conforming to design rules and interfaces. "Modularity in design should tremendously boost the rate of innovation in many industries as it did in the computer industry," say Baldwin and

Clark. But moving to modularity alters the balance of prototyping power. "No longer will assemblers control the final product: suppliers of key modules will gain leverage and even take on responsibility for design rules," they predict. "Companies will compete either by specifying the dominant design rules (as Microsoft does) or by producing excellent modules. . . . Leaders in a modular industry will control less, so they will have to watch the competitive environment closely for opportunities to link up with other module makers."

The unit of prototyping thus shifts from the product or component to the module. As long as the module does its job according to the rules, it's welcome. Modeling in modular industries will be radically different from modeling in industries that call for collaboration and transparency beyond the interfaces.

These examples illustrate the overriding principle about the future of the modeling marketplace: relationships are what matter most. Even in the most competitive marketplace, critical distinctions exist between cooperative relationships, collaborative relationships, and transactional relationships. Which modeling relationships are likeliest to create the most value at the most reasonable cost?

The content of relationships—especially those based on the creation and exchange of value—is behavior. The question that matters isn't, Which models will create the best value? It's, Which modeling behaviors and relationships will create the best value?

> **The idea that technology will promote a convergence of "best practices" is wrongheaded.**

Consequently, the idea that technology will promote a grand convergence of "best practices" is wrongheaded. On the contrary: instead of smoothing over cultural distinctions, the proliferation of modeling media is likely to amplify and sharpen them. Opportunities for meaningful differentiation will grow, not diminish. Cultural and behavioral issues will become even more decisive than they are today. Diversity in innovation infrastructures will increase. The pace of business evolution will accelerate.

Design versus Darwin

In business, evolution is no longer just a metaphor; it has evolved into an innovation methodology. Malcolm Smith, a cofounder of Palo Alto Products International, a leading industrial-design firm, asserts that his company's innovation algorithm is, "Design it. Build it. Test it. Repeat 5 million times." Clever. But after how many iterations is something that has been designed transformed into something that evolved? The distinction isn't semantic. The more cycles and iterations a prototype undergoes, the less it reflects underlying plans and the more it reflects evolutionary responses to environmental pressures. As computing costs plummet and processing speeds accelerate, the ability to evolve iterations is becoming an important complement to designing iterations. In some areas of design—notably, software algorithms and computational fluid dynamics—evolutionary iterations are already a substitute for design iterations. The algorithms, sometimes called genetic algorithms, literally take on a life of their own.

Today, CAD/CAE stands for Computer-Aided Design and Computer-Aided Engineering. Tomorrow, the same acronyms may stand for Computer-Aided Darwinism and Computer-Aided Evolution. According to leading computer researchers, software engineering will acquire an evolutionary cast. State-of-the-art software breakthroughs won't merely be written, they will be grown. Instead of just composing programs, tomorrow's programmers will also breed them. "I think of it as being like a software brewmaster," says Disney Imagineer Danny Hillis, an evolutionary-programming pioneer. "This [approach] gives us the potential to get computers to do things that are more complicated than we understand."

By treating the computer as an environment and the software as organisms that can be made to evolve over time, a new discipline of software sociobiology emerges. Instead of writing a bug-free set of instructions, the designer-programmer imbues the program with certain traits—software "genes"—and then defines the environmental constraints within which the resulting organisms will evolve. Research and real-world testing confirm that evolved pro-

grams can be far more efficient—and less expensive to develop—
than handcrafted ones. Hillis, John Holland, and other leading
genetic programmers are confident that
the evolutionary-software genre will absorb an ever-larger portion of the calculations now performed by traditional
computer-aided design/computer-aided
engineering algorithms. Increasingly, in-
novation design will feel more like editing a movie than shooting
one. Innovation will be more about selection pressures than creation anxieties.

Increasingly, innovation design will feel more like editing a movie than shooting one.

"We all find programming very frustrating," Hillis has remarked.
"The dream is to have this little world where we can evolve programs to do something we want. We punish them and wipe them
out if they don't do what we want, and after a few hundred thousand generations or so, we have a program that ideally behaves the
way we like. We can play god—but all we have to do is define the
puzzle for them. We don't have to be smart enough to figure out
the way they solve the puzzle."

As digital infrastructures such as the Internet become ever faster
and cheaper, our ability to evolve robust and sophisticated software
organisms increases. A fault-tolerant network may be more cost-
effectively evolved than designed; so might a microprocessor or a
financial instrument. Pharmaceutical companies such as Merck
and Pfizer are exploring how to evolve new medicines using
genetic algorithms with the same intensity with which they've
invested in biotechnology. In many respects, evolution is the ultimate prototyping and simulation methodology. Evolution's power
and versatility are inarguable; its ability to innovate and surprise is
overwhelming.

Richard Dawkins, a leading evolutionary theorist, once described
using a simple genetic algorithm to try to grow a forest of virtual
trees on his old Apple computer:

> When I wrote the program, I never thought it would evolve anything
> more than a variety of treelike shapes. . . . Nothing in my biologist's

intuition, nothing in my 20 years' experience of programming com-
puters, and nothing in my wildest dreams, prepared me for what
actually emerged on the screen. I can't remember exactly when in the
sequence it first began to dawn on me that an evolved resemblance to
something like an insect was possible. With a wild surmise, I began
to breed, generation after generation, from whichever child looked
most like an insect. My incredulity grew in parallel with the evolving
resemblance. . . . Admittedly [my bugs had] eight legs like a spider,
instead of six like an insect, but even so! I still cannot conceal from
you my feeling of exultation as I first watched these exquisite crea-
tures emerging before my eyes. I distinctly heard the triumphal open-
ing chords of "Also Sprach Zarathustra" (the 2001 theme) in my
mind. I couldn't eat, and that night "my" insects swarmed behind
my eyelids as I tried to sleep.

This was the experience of a creative evolutionary zoologist playing with crude, handmade artificial-life software on a first-generation personal computer. The simplest algorithm generated a vivarium of insectoid innovation that beggared Dawkins's imagination. The technologies of today and tomorrow are orders of magnitude faster and more powerful. The software genomes enabling new breeds of innovation are far richer and more complex.

Imagine the evolutionary innovations a software designer might breed on the Internet, or the aerodynamic shapes an automotive engineer might evolve on a networked workstation, or the derivative instruments an investment banking quant might breed for a financial-services marketplace, or the array of pricing schemes a yield-management expert could evolve for a global logistics firm, or the product extensions a brand manager could grow for niche market seg-

Innovators will be the stewards of artificial life forms.

ments. The prototypes for these speculative innovations would once have been designed; now the economics of evolutionary prototyping allow them to be bred and culled. The best of breed can be profitably harvested to become software "seeds" for a new species of innovation.

As Hillis suggested, innovators will play god. Or, more accurately, they will be the stewards of artificial life forms that serve as vibrant innovation populations. Much as market mechanisms will play a greater role in managing prototype proliferation, quantitative population-genetics analysis will become essential to managing these new breeds of simulation. Just as innovators looked to statisticians and operations researchers to manage complex business-logistics algorithms in the 1950s and 1960s, the millennium will create a demand for population geneticists and other mathematically oriented managers. The creative tensions between innovators who design and innovators who evolve will likely result in breakthroughs in products, services, and their yet-to-be-anticipated hybrids.

Modeling Models

In the future, innovators may want to prototype how they prototype or to play simulation games that give them insight into how they simulate. Just as prototypes and simulations speak volumes about how organizations innovate, the ability to model the innovation process gives innovators the power to view their assumptions and taboos from a different perspective. How valuable will that perspective prove to be in the marketplace?

Companies like Boeing, General Motors, and Merrill Lynch already simulate the business implications of integrating products and services. For example, General Motors, General Electric, and Boeing must align their manufacturing and production capabilities with their leasing arms. The "unit of prototyping" is shifting from individual products, processes, and services toward hybrids. Those hybrids in turn raise fundamental questions about the organization itself. How might the firm be reorganized to exploit tomorrow's innovative hybrids? How should

> The unit of prototyping is shifting from individual products and services to hybrids.

those reorganizations be modeled? What new organizational forms might evolve? Indeed, should simulations be run with the expectation that the organization needs to be redesigned, that it must evolve?

Concluding a book with these questions might seem unfair. However, these are exactly the questions that tomorrow's CEOs, investors, and entrepreneurs will be forced to ask as their markets grow ever more competitive and their rivals more innovative. Their best hope for answering these questions, or coping with their implications, is to build or grow the right models and play with them seriously.

The Best Policy

A time-pressed innovator hungry to benefit from serious play might prefer a book entitled *The Seven Habits of Highly Effective Simulators* or *The One-Minute Modeler*. This is not that book. But there are some ways to boost the odds of superior returns on investments in serious play. This user's guide is presented in that spirit.

These rules offer actionable insights. They work. They're heuristics for managers and project leaders who want to proceed with awareness and a sense of urgency. They reflect the singular reality that dominates the enterprise: what organizations do is more important that what they say. Managerial choices speak more eloquently than mission statements.

But be warned: these rules pose a serious problem. They demand honesty. They force people to look at themselves through their prototypes. All too frequently, firms don't like what they see. So they ignore it, suppress it, destroy it, or dismiss it as an aberration.

Without question, the biggest challenge that serious play poses isn't a matter of technology or intelligence or information; it's dishonesty. Will Rogers once observed that "it's not what you don't know that hurts you; it's what you do know that ain't so." Actually, what hurts most organizations is willful refusal to credit what their models are saying about what they know and don't know. How models are used and abused reveals a great deal about the character of a corporation. Healthy prototyping cultures discourage self-deception and misrepresentation.

Dishonest play can lead to dishonest designs. Alan Stokes and Kirsten Kite of Rensselaer Polytechnic offer an example in *Caveman in the Cockpit: Evolutionary Psychology and Human Factors*:

> Consider the federal rule that airliner cabins be designed such that they can be evacuated in 90 seconds. For many years (and continuing) this capability was demonstrated in manufacturer and FAA tests, the "passengers" emptied from the cabin in a smooth orderly way in 90 seconds or less. Analysis showed that regularity and efficiency is assured by waiting, deference and turn-taking—in short by individuals acting for the welfare of the group in the simulated emergency. But is this descriptive of the human psychological fit, the real output of hot cognition when the prime biological directive "Survive!" is invoked?

> Put another way, are we the descendants of folk who deferred to others and waited their turns in emergencies, or are we blessed with the genes of those who dropped everything and ran like blazes? When a cabin evacuation experiment was conducted at Cranfield U.K., the mere chance of collecting about $8 (for the first 30 out) was enough to create bedlam in the cabin-jammed exits. . . . This is hardly an unexpected outcome of esoteric genetic theory, yet we had to wait for a fatal 737 cabin fire at Manchester Airport before such a study was funded, revealing the 90-second rule for what it is—a polite fiction rooted more in corporate wishful thinking than in the operational "specs" for homo sapiens.

In how many other organizations do polite fictions riddle proposed innovations? Would a simple simulation invalidate the assumptions that preserve them? The rules in this user's guide won't work very well for managers and innovators who depend on polite fictions.

1. Ask, Who benefits?

The first and most important question that must be asked isn't What is this model for? or What are we really trying to do? It's, Who benefits?

The corollary question is equally penetrating: Who stands to lose if a useful prototype or simulation is created? One clear lesson of serious play is that models can transform the balance of power in the enterprise. The "innocent" modification of a simulation can have shockingly dramatic consequences. A brave new prototype may inspire new levels of interdepartmental collaboration or cross-functional fratricide. Who is responsible for anticipating those consequences? Who is asking—and answering—Who benefits?

Whenever aspiring innovators believe that a rapid prototype is in order, an equally rapid analysis of who stands to win or lose becomes paramount. For instance, a prototype customer-support Internet site might be brilliantly designed, but if it threatens the company's sales force it becomes an exercise in the politics of polarization. Conversely, a simulation demo supporting the chairman's upcoming keynote speech at a prestigious industry conference may have disproportionate impact. Prototypes always have political ramifications; prototypes let people play politics.

Asking who really benefits from serious play—Customers? Clients? The CEO? Marketing? Manufacturing? Particular managers?—helps shape how the model will develop. So will asking, Who does this model threaten? The answers to these questions pinpoint allies who might be enlisted and rivals who may need to be mollified—or neutralized—as the model evolves. This awareness is essential if serious play is to be elevated from an intellectual exercise to a productive behavior.

But—as innovators from Microsoft to IKEA to BMW can confirm—as a prototype evolves, so do the answers to the question, Who benefits? An innovation initially intended to support the core business can mutate into a form that would undermine it, and vice versa. Who benefits? is a dynamic question that must be asked and answered at every significant stage of a prototype's development. Do a benefits/threat map or a spreadsheet.

Otherwise, the outcome is likely to be one more instance of "Pyrrhic prototyping": prototypes that deliver valuable insights but are ignored because their costs outweigh the benefits.

2. Decide what the main paybacks should be and measure them. Rigorously.

Is the purpose of the prototype to save time? Save money? Persuade customers? Actively involve suppliers? Generate new knowledge? Demonstrate feasibility? Communicate the nature of the innovation to top management? Any of these might be a worthy goal. But most organizations do a cavalier or lazy job of explicitly measuring the desired benefits of their prototyping process.

Successful prototyping cultures link changes in their models to metrics they consider critical. On New Zealand's America's Cup champion yacht *Black Magic*, for instance, design modifications and iterations were tested in computer simulations and in races with its not-quite-identical twin. IKEA calculates the manufacturing costs of furniture prototypes down to a fraction of a *krone*: Cost-to-build is explicitly identified as IKEA's single most important criterion. At Steelcase's New York prototyping showroom, customers' length of stay was measured as a gauge of their interest in Steelcase systems. World-class companies that differentiate themselves with design are careful to use mission-critical measures in their prototyping processes.

The quest for ever-faster speed-to-market has become a design imperative. Increasingly, organizations are using computer-aided design and engineering systems to accelerate product development. Service companies too are using digital systems to increase their rate of innovation. Rapid prototyping and simulation has become a primary reason for adopting new modeling media. Speed is significant and fairly straightforward to measure. But speed-to-market metrics must be handled with great care. An accelerated rate of innovation can outstrip the customer's ability to absorb it.

A meaningful prototyping metric must align a prototype's value within the firm to its potential value in the marketplace. In other words, prototyping payback metrics should help link how the organization creates the prototype to how the customer ultimately values it. Customer paybacks matter as much as internal payback metrics.

3. Fail early and often.

Stefan Thomke of Harvard Business School persuasively argues that organizations that aggressively use models to identify design conflicts early in the development process achieve significant economies of innovation over firms that do not. This approach, which Thomke calls "frontloading," can, he asserts, "reduce development time and cost and thus free up resources to be more innovative in the marketplace."

In effect, rapid prototyping means rapid failures. This is no mere paradox but a vital design principle which world-class companies are embracing. Digital modeling media encourage the "frontloading" of failures. Industry leaders such as Toyota, Boeing, and DaimlerChrysler have become adept at generating low-fidelity simulations that offer accelerated insight into key design conflicts. Frontloading captures these critical conflicts as early in the development cycle as possible. Given that the later such design conflicts are corrected, the more expensive they are to fix, this "fail-early" approach is becoming an imperative.

Thomke's work with German automaker BMW revealed that computer simulation of crashworthiness simultaneously accelerated the pace of design iterations and reduced their costs. Reviews of ninety-one design iterations showed that developers were able to improve the side-impact crashworthiness of a BMW sedan by about 30 percent—"an accomplishment that would have been unlikely with a few lengthy problem-solving cycles using physical prototypes only," according to Thomke.

The same "design-for-failure" ethic is applicable to the simulation of experiences. Financial-service firms that create and sell high-risk derivative instruments are increasingly investing computer time and human creativity in scenarios and simulations in which their synthetics fail. Given the volatility of global markets, these stress tests enhance the credibility—and salability—of the innovative instruments that survive. The key point is that financial innovators, like their automotive counterparts, use digital technologies to frontload their failures.

4. Manage a diversified prototype portfolio.

Nobel Prize–winning economics research affirms that rational diversification is essential to managing financial risks and opportunities. Many prototyping cultures, however, ignore this principle and overinvest in a single mode of representation as a "monopoly" modeling medium. This approach badly hurt Detroit's automakers, for example, when clay was the dominant prototyping medium for styling.

The dominant modeling medium, whatever it may be, inherently distorts prototyping perceptions and processes. Industrial designer John Rheinfrank stresses that innovative organizations should create multiple models and prototypes of their innovations. "New-product sketches should be turned into storyboards that describe an experience," he asserts, "and storyboards should be used to create both 'looks-like' and 'works-like' physical prototypes of a product." The goal is to encourage interplay and dialogue between the various representations.

Interplay of this kind has proven particularly important in the realm of virtual prototyping. Creating design dialogues between CAD representations of products and their physical counterparts has proven extraordinarily valuable for iterating improvements and enhancements. The architect Frank Gehry, for example, uses both CATIA-driven virtual models and physical models of various scales as a way to better visualize the ultimate form and finish of his buildings. "I need multiple models to really see what I am doing," he explains.

It is understandable that firms want standardized CAD products or simulation software, but it's equally clear why a host of less formal prototypes inevitably spring up around the standard. Some are symbionts; others are parasites. Sketches, mathematical equations, bootleg software, and even customized versions of the standard are all ultimately part of a diversified prototyping portfolio. Managers ought to pay special attention to which parts of these diversified portfolios generate the most interesting questions, and the most useful answers. Design discussions around whiteboard sketches may be far more influential than the more technically explicit conversations that go on around a computer screen.

The most important mental shift that must occur is for innovators to stop managing "the Prototype" and start managing the multiple models that feed into an innovation. In many respects, successful innovators are as much de facto "modeling mutual-fund managers" as they are "product champions."

5. Commit to a migration path. Honor that commitment.

Prototypes will not command time and respect unless they are seen to have a direct impact on the final product or service. Fortune 100 giants and Silicon Valley start-ups alike have spun their wheels on prototypes that could never have been incorporated into their production and distribution infrastructures.

Unless managers indicate unambiguously how a proposed model or simulation will ultimately be mapped into the ultimate product, it is likely to be a waste of time. The more tightly-coupled a prototype is to the ultimate product, the more likely it is to inspire innovative thinking. When computer-aided design software was integrated with computer-aided engineering and computer-aided manufacturing software, its perceived value skyrocketed in organizations ranging from BMW to Caterpillar to Canon. When leveraged buyout mavens Kohlberg Kravis Roberts discovered in the

1980s that the software spreadsheets they had originally used to analyze a desirable enterprise could also be used to help run that firm, the role of analyses assumed a richer focus and intensity.

At some organizations, such as CalTech's Jet Propulsion Laboratory, a few partial differential equations can have a greater impact on a design than a week's worth of supercomputer time. Conversely, the outputs of some organizations' model shops and pilot plants are routinely ignored when it comes time to make the hard investment decisions. While it might seem excessive for every prototype or simulation to have its own mission statement, it is the essence of good management to explain how the benefits of the proposed model will unambiguously outweigh the costs of building and using it.

6. A prototype should be an invitation to play.

You know you have a successful prototype when people who see it make useful suggestions about how it can be improved. Successful innovators don't use their prototypes to persuade the right people; their prototypes enable the right people to persuade themselves. If your most important boss, client, or supplier can't play creatively with your prototype, you have made a serious design error.

Prototypes should lure people into innovative games of "What-if?" Prototypes should turn customers, clients, colleagues, and vendors into collaborators. When Peter Schwartz was at Royal Dutch/Shell, his ability to turn a software spreadsheet into a theater for future pricing scenarios changed conversation in the firm. The ability of Dell's customers to prototype configurations of personal computers before they buy has proven extraordinarily valuable. The successful efforts of Boeing, Toyota, and Chrysler to bring their best suppliers onto their design teams have fundamentally changed their innovation economics.

Anything an organization can do to make its prototypes more transparent and accessible to its core constituencies should be ruthlessly explored. Instead of investing time eliciting customer requirements to build prototypes, innovators should be using prototypes to elicit requirements. Prototypes should be managed as media to track people in the act of thinking out loud—and then in playing out those thoughts with the prototype. Serious play becomes peer review.

Detroit's clay sculptures didn't invite editing and enhancement; computer-aided design software does. Does that capacity create the risk that amateurs will ruin the representations of experts? Of course. That's why such invitations should emphasize play. But to treat prototypes as the end of a dialogue, rather than a chance to create new dialogues, represents a tremendous waste of resources. When the right customers, vendors, and colleagues play with your prototypes, errors can be captured before they become obstacles; serendipity becomes a colleague. The more flexible and dynamic the prototypes, the more flexible and dynamic the play—and the greater the opportunities for profitable innovation. Design the invitations accordingly.

7. Create markets around the prototypes.

Playing with prototypes may be fun and informative, but should it be free? Should people pay to play? Microsoft enjoyed the equivalent of a $1-billion subsidy from its best customers and independent developers while developing its Windows 95 operating system. In fact, Microsoft's strategic management of its beta sites has consistently been a source of competitive advantage for the world's most profitable software company.

Such a lavish subsidy is a prerogative of market leaders, but it speaks volumes about how innovators can get their best customers and suppliers to subsidize innovation efforts. As the Microsoft example illustrates, there are currencies other than money. Valuable subsidies come in many forms, including information. For example, investment banks' financial engineers can learn a great deal from institutional investors about the risk-management and investment opportunity issues they confront. Even if the information doesn't give rise to a new financial instrument, the knowledge gleaned can be invaluable. Even noting which features of a product or service are actually used, and which ignored, can be significant. As former Netscape president James Barksdale observed during his company's heyday, if a product Netscape posted on the Web was successful, terrific; if not, it was just market research.

Any organization not using its prototypes and simulations to help create subsidies—measurable in time, money, or human capital—from its key constituencies is mismanaging its prototyping potential. But the significance of a prototyping marketplace goes beyond subsidies. As Boeing has demonstrated, organizations can also use market forces to help make their prototyping processes more cost-effective. Boeing's use of a "weight economy" rather than a "weight budget" to manage development of its 777 jet affirms that the economics of prototyping invite fundamental rethinking. A prototype need not be something a department simply pays for; instead it could be treated as part of an economy. Perhaps there should be costs associated with making changes at certain times in the prototyping process. Customers and suppliers who participate in the design process could receive certain benefits when the innovation is finally launched. Other parts of the organization could have budgets for helping shape the prototype; for example, customer support might pay engineering to make a costly modification in the design that will ultimately reduce the cost of after-market customer service. In sum, cost-effectively managing the prototyping process now requires innovators to design internal markets with the same creativity they bring to their innovations.

8. Encourage role playing.

Prototypes and simulations lend themselves to being treated as theaters in which participants in the innovation process assume different roles. The most obvious role, of course, is that of customer. Citigroup, Merrill Lynch, Morgan Stanley Dean Witter, and British Air have all dramatically changed their customer-service infrastructures in response to designers' experience with simulations of the support systems they proposed to build. Simulating the customer experience—not just observing it through a one-way mirror—before inflicting it on the marketplace can be an influential bit of prototyping theater.

As cross-functional, cross-disciplinary teams become a dominant medium for managing innovation, prototypes and simulations can promote awareness and empathy between collaborators. It is a useful exercise for a design engineer to simulate the constraints imposed on manufacturing engineers and for an insurance agent to experience those confronting the claims adjuster. It is as edifying for an airline pilot to simulate the task of an air-traffic controller as it is for the air-traffic controller to sit in the cockpit of a Boeing 737 simulator. Conversations during such role playing tend to be insightful. It is not enough for prototypes to be the shared medium of communication and collaboration in the innovation process; they should also be a tool for people to step outside of their everyday roles. Why? Because seeing the value of a prototype or simulation requires viewing it from a different perspective. The surest way to view something from another perspective is to adopt another perspective. The goal should not be cross-functional fluency but cross-functional awareness. It is often difficult to get managers to take this sort of exercise seriously, but videotaping such performances boosts the odds that people will think twice before acting too glib.

9. Determine the points of diminishing returns.

The computational costs of rapid prototyping are shriveling. Digital technologies dramatically accelerate the ability to iterate. Opportunities to "productively waste" development cycles are expanding. This embarrassment of prototyping riches creates an awkward management puzzle: just how much of a good thing is too much?

An organization that has just acquired new modeling media frequently behaves like a writer with a new word processor: both endlessly futz around simply because they can. Instead of a vehicle for value creation, these discount technologies can easily become an effective way to waste time and effort. What makes the thirtieth iteration of a prototype more valuable than the twenty-fifth? If a spreadsheet scenario can be stress-tested a thousand different ways, should it be?

A common pathology is to keep iterating right up to the last possible moment. The deadline becomes the criterion for conclusion. The platitude, "we can always make things a little better" overwhelms the question of whether the benefits of marginal improvement genuinely outweigh its costs. At a minimum, there are opportunity costs.

In an environment of computational abundance, firms create value by creating constraints. These constraints can be cultural or economic, objective or subjective. At Royal Dutch/Shell, the iterations cease when conversations become echoes of prior conversations rather than sources of insight. Procter & Gamble's one-page memo is a superb example of a culturally and managerial imposed constraint on communication. Boeing and Caterpillar discourage CAD/CAE iteration for the sake of iteration by tracking why iterations are made. The goal is not to suppress the "just-one-more" design philosophy but to insist on intellectual rigor.

Iterating for the sake of iterating is an indulgence; iterating for the sake of curiosity is research; iterating for the sake of value creation demands criteria to guarantee that the benefits outweigh the costs. In effect, identifying a point of diminishing returns acknowledges the difference between innovation as a design imperative and innovation as a business practice.

10. Record and review relentlessly and rigorously.

There are few world-class athletes who do not videotape and review their practices. Videotaping performances has also become a critical part of managing rehearsals for theater, television, and films, asserts Julie Taymor, who staged *The Lion King* on Broadway. Microsoft performs postmortems on its software products after they've completed their internal development cycles. The U.S. national-security establishment considers "hot washup"—the review process after a war game—critical for learning the lessons of the virtual conflict.

The best and most honest way for people to learn from their prototypes is to examine how they built and used them. Digital modeling media enable companies to create "introspection infrastructures" that give people the tools to record how they and their prototypes coevolved over time: to videotape, audiotape, and otherwise record human interactions surrounding prototypes and to link them to changes in the prototypes themselves. To make the most of these tools, organizations have to commit to reviewing "Lessons Learned," "Worst Mistakes," and "Mistakes Avoided."

Design leaders like Walt Disney and IDEO also videotape and ethnographically analyze how their product prototypes work in the field. How do clients and customers actually experience the prototypes (as opposed to merely listening to what they have to say about them)? How do customers modify these experiences when given the chance?

Such information should be reviewed in the context of previous lessons. Are interactions around models and simulations changing for the better or the worse? Instead of relying on textual reports or static PowerPoint presentations, organizations will increasingly have the option of an interactive animated review of their innovation process. They will be able to see how their design and marketing conversations changed as the prototypes changed. They will be able to assess their past conversations and design decisions in the context of the choices that were actually made.

The key to learning is to invest in an introspection infrastructure that dramatically expands the bandwidths of memory and context. If world-class basketball teams and soccer teams use multiperspective videotapes to review their practices and performances, why shouldn't world-class business teams? If world-class sports teams use computer analysis to see which formations and plays enjoy the greatest chance of success in given situations, why shouldn't world-class innovators? The answer is simple: they should, they must, and they will. The ability to review the prototyping process will become as valuable as the prototyping process itself in the creation of new value.

BIBLIOGRAPHY

Abbott, Walter F. "Shadow Juries: A Checklist of Practical Decisions." *Trial Diplomacy Journal* (Fall 1987): 30–35.

Abt, Clark C. *Serious Games.* New York: Viking Press, 1971.

Ackoff, Russel L. *Ackoff's Best: His Classic Writings on Management.* New York: John Wiley & Sons, 1999.

Adler, Paul S., and Terry A. Winograd. *Usability: Turning Technology into Tools.* New York: Oxford University Press, 1992.

Allen, Thomas B. *War Games: The Secret World of the Creators, Players and Policy Makers Rehearsing World War III Today.* New York: McGraw-Hill, 1987.

Anderson, Richard E. "Strategic Integration: How John Deere Did It." *Journal of Business Strategy* 13, no. 4 (1992): 21–26.

Andlinger, G.R. "Business Games Play One!" *Harvard Business Review* (1958): 115–125.

Anfuso, Dawn. "Turning Business Into a Game." *Personnel Journal* 74, no. 3 (1995): 50–56.

Ang, James, and Jess Chua. "Corporate Planning Models That Failed." *Managerial Planning*, July/August 1980, 3–11.

Ardhaldjian, Raffy, and Mike Fahner. "Using Simulation in the Business Process Reengineering Effort." *Industrial Engineering* 26, no. 7 (1994): 60–61.

Aris, Rutherford. *Mathematical Modeling Techniques.* New York: Dover Publications, 1994.

Ashton, Alison Hubbard, Robert H. Ashton, and Mary N. Davis. "White-Collar Robotics: Leveraging Managerial Decision Making." *California Management Review* 37, no. 1 (Fall 1994): 83–109.

Axelrod, Robert. *The Complexity of Cooperation: Agent-based Models of Competition and Collaboration.* Princeton, N.J.: Princeton University Press, 1997.

Bakken, Bent, Janet Gould, and Daniel Kim. "Experimentation in Learning Organizations: A Management Flight Simulator Approach." *European Journal of Operational Research* 59 (1992): 167–182.

Bankes, Steve. "Exploratory Modeling for Policy Analysis." RAND paper, Santa Monica, Calif., 1993.

Bankes, Steven C. "Computational Experiments and Exploratory Modeling." RAND paper, Santa Monica, Calif., 1994.

Bankes, Steven, and James Gillogly. "Exploratory Modeling: Search Through Spaces of Computational Experiments." RAND paper, Santa Monica, Calif., 1994.

——. "Validation of Exploratory Modeling." RAND paper, Santa Monica, Calif., 1994.

Banks, Jerry, ed. *Handbook of Simulation: Principles, Methodology, Advances, Applications and Practice.* New York: John Wiley & Sons, 1998.

Barkow, Jerome H., Leda Cosmides, and John Tooby. *The Adapted Mind: Evolutionary Psychology and the Generation of Culture.* New York: Oxford University Press, 1992.

Barnard, Chester I. *The Functions of the Executive.* Cambridge: Harvard University Press, 1937.

Barret, Philip. "Spreadsheets: Give the Workhorse Plenty of Rein." *Accountancy*, March 1985, 161–162.

Bayley, Stephen. *Design Heroes: Harley Earl.* New York: Grafton Press, 1990.

Bedikian, Mary A., and Jerome D. Hill. "The Ultimate Power of Persuasion: Using the Mock Trial to Enhance Litigation Strategy." *Michigan Bar Journal*, October 1993, 1046ff.

Bekey, George A. "Models of Reality: Some Reflections of the Art and Science of Simulation." *Simulation*, November 1977, 161–164.

Bergstrom, Robin Yale. "Okay, So You're Not Having Fun Yet, But Who Said You Would?" *Production*, May 1994, 16.

Bielecki, Witold Tomasz. "Turning Business Simulation into a Decision Support System." *Journal of Management Development* 12, no. 3 (1993): 60–64.

Bijker, Wiebe E., Thomas P. Hughes, and Trevor Pinch, eds. *The Social Construction of Technological Systems: New Directions in the Sociology and History of Technology.* Cambridge: MIT Press, 1987.

Bloomfield, Lincoln P. "Reflections on Gaming." *Orbis* 27 (Winter 1984): 783–790.

Bodily, Samuel. "Spreadsheet Modeling as a Stepping Stone." *Interfaces* 16, no. 5 (1986): 34–52.

Boehm, Barry, T. Gray, and T. Seewaldt. "Prototyping versus Specifying: A Multiproject Experiment." *IEEE Transactions on Software Engineering*, May 1984, 290–302.

Boettinger, Henry. *Moving Mountains: The Art and Craft of Letting Others See Things Your Way*. New York: Macmillan, 1969.

Bonini, Charles. "Computers, Modelling, and Management Education." *California Management Review* VXXI, no. 2 (1978): 47–55.

Booker, Ellis. "Have You Driven a Simulated Ford Lately?" *Computerworld* 28, no. 27 (1994): 76.

Bouden, J. B., and E. S. Buffa. "Corporate Models: On Line, Realtime Systems." *Harvard Business Review*, July–August 1970, 65–83.

Bowen, H. Kent, et al., eds. *The Perpetual Enterprise Machine: Seven Keys to Corporate Renewal through Successful Product and Process Development*. New York: Oxford University Press, 1994.

Brady, R. "Computers in Top-Level Decision-Making." *Harvard Business Review*, July–August 1967, 67–76.

Brand, Stewart. *How Buildings Learn*. New York: Viking Penguin, 1994.

Brenner, Reuven. "Extracting Sunbeams Out of Cucumbers: What Is Bad Social Science, and Why Is It Practiced?" *Queen's Quarterly* 98, no. 3 (1991): 519–553.

Brewer, Bill. "A Company-Wide Business Game." *Journal of European Industrial Training* 16, no. 1 (1992): i–ii.

Brewer, Garry D., and Martin Shubik. *The War Game: A Critique of Military Problem Solving*. Cambridge: Harvard University Press, 1979.

Brook, Peter. *There Are No Secrets: Thoughts on Acting and Theatre*. London: Methuen, 1993.

Brooks, Frederick P., Jr. *The Mythical Man Month: Essays on Software Engineering*. Reading, Mass.: Addison-Wesley, 1995. (Anniversary Edition.)

Brown, Kenneth. *Inventors at Work*. Redmond, Wash.: Microsoft Press, 1988.

Brown, Shona L., and Kathleen M. Eisenhardt. *Competing on the Edge: Strategy as Structured Chaos*. Boston: Harvard Business School Press, 1998.

Bruce, Margaret, and Birgiy H. Jevnaker, eds. *Management of Design Alliances: Sustaining Competitive Advantage*. New York: John Wiley & Sons, 1997.

Bruner, Jerome. *Actual Minds, Possible Worlds*. Cambridge: Harvard University Press, 1996.

————. *The Culture of Education*. Cambridge: Harvard University Press, 1996.

————. *On Knowing: Essays for the Left Hand*. Cambridge: Harvard University Press, 1979.

Bucciarelli, Louis L. *Designing Engineers*. Cambridge: MIT Press, 1994.

Buchanan, Richard, and Victor Margolin, eds. *Discovering Design: Explorations in Design Studies*. Chicago: University of Chicago Press, 1995.

Bunch, John. "The Storyboard Strategy." *Training and Development*, July 1991, 69–71.

Burch, John G., Jr. "Business Games and Simulation Techniques." *Management Accounting* 51 (December 1969): 49–52.

Busby, Scott. "Imagining Movies." *Premiere*, July 1989, 68–71.

Bush, Akiko, ed. *Design for Sports: The Cult of Performance*. New York: Princeton Architectural Press, 1989.

Butler, Keith A. "Usability Engineering Turns 10." *Interactions* (January 1996): 59–75.

Byrne, John A. "Virtual Management: Computer Models May Give Execs Previews of How Decisions Pan Out." *Business Week*, 21 September 1988.

Campbell-Kelly, Martin, and William Aspray. *Computer: A History of the Information Machine*. New York: Basic Books, 1996.

Canemaker, John. *Before the Animation Begins*. New York: Hyperion, 1996.

Caplan, Ralph. *By Design: Why There Are No Locks on the Bathroom Doors in the Hotel Louis XIV and Other Object Lessons*. New York: St. Martin's Press, 1980.

Carley, Kathleen M., and Michael J. Prietula, eds. *Computational Organization Theory*. Hillsdale, N.J.: Lawrence Erlbaum Associates, 1994.

Carroll, John M., ed. *Scenario-Based Design: Envisioning Work and Technology in System Development*. New York: John Wiley & Sons, 1995.

Casti, John L. *Would-be Worlds: How Simulation Is Changing the Frontiers of Science*. New York: John Wiley & Sons, 1997.

Ceruzzi, Paul E. *Beyond the Limits: Flight Enters the Computer Age*. Cambridge: MIT Press, 1989.

Christopher, Elizabeth. "The Day I Learned to Play Games." *Simulation & Gaming*, December 1995, 420.

Chussil, Mark J. "Competitive Intelligence Goes to War: CI, the War College, and Competitive Success." *Competitive Intelligence Review* 7, no. 3 (1996): 56–69.

Clarke, S., and A. M. Tobias. "Complexity in Corporate Modelling: A Review." *Business History* 37, no. 2 (1995): 17–44.

Clifford, Robert. "Mock Trials Offer Virtual Reality." *The National Law Journal*, 27 February 1995, B8.

Clippinger, John H. "Visualization of Knowledge: Building and Using Intangible Assets Digitally." *Planning Review* 23, no. 6 (1995): 28–31.

Clurman, Harold. "Mysterious Rites of Rehearsal." *New York Times Magazine*, 5 March 1961, 54–67.

Cole, Susan Letzler. Directors in Rehearsal. New York: Routledge, 1992.

Coles, Robert. *Doing Documentary Work.* New York: Oxford University Press, 1997.

Collins, Michael J. "Benchmarking with Simulation: How It Can Help Your Production," *Production* 107, no. 7 (1995): 50–52.

Cragg, Paul B., and Malcolm King. "Spreadsheet Modelling Abuse: An Opportunity for OR?" *Journal of Operational Research* 44, no. 8 (1993): 743–752.

Crick, Francis. *What Mad Pursuit?* New York: Basic Books, 1989.

Crouch, Tom. *The Bishop's Boys: A Life of Wilbur and Orville Wright.* New York: W.W. Norton, 1989.

Cusumano, Michael, and Richard Selby. "Microsoft's Weaknesses in Software Development." *American Programmer* (October 1997): 8–13.

Cziko, Gary. *Without Miracles: Universal Selection Theory and the Second Darwinian Revolution.* Cambridge: MIT Press, 1995.

Darnton, Nina. "Choosing a Climax." *NY Post*, 9 October 1987.

———. "Lights! Camera! Bring in the Test Audience." *NY Post*, 18 September 1987.

Davis, Paul K., and Donald Blumenthal. "The Base of Sand Problem: A White Paper on the State of Military Combat Modeling." RAND paper, Santa Monica, Calif., 1991.

Dawkins, M. S., T. R. Halliday, and R. Dawkins, eds. *The Tinbergen Legacy.* New York: Chapman & Hall, 1991.

Dawkins, Richard. *The Blind Watchmaker: Why the Evidence of Evolution Reveals a Universe without Design.* New York: W.W. Norton, 1987.

de Geus, Arie. *The Living Company: Habits for Survival in a Turbulent Business Environment.* Boston: Harvard Business School Press, 1997.

DeGrace, Peter, and Leslie Hulet Stahl. *Wicked Problems, Righteous Solutions: A Catalogue of Modern Software Engineering Paradigms.* Englewood Cliffs, N.J.: Prentice-Hall, 1990.

DeMarco, Tom. *A Novel about Project Management.* New York: Dorset House, 1998.

———. *Why Does Software Cost So Much and Other Puzzles of the Information Age.* New York: Dorset House, 1995.

DeMarco, Tom, and Timothy Lister. *Peopleware: Productive Projects and Teams.* New York: Dorset House, 1987.

Dembo, Ron S., and Andrew Freeman. *Seeing Tomorrow: Rewriting the Rules of Risk.* New York: John Wiley & Sons, 1998.

Denning, Peter J., and Robert M. Metcalfe, eds. *Beyond Calculation: The Next Fifty Years of Computing.* New York: Springer-Verlag, 1997.

Devine, Foy R. "What Goes On in the Jury Room?" *Trial Lawyers Quarterly,* Spring 1988, 7–13.

Dewsbury, Donald A., ed. *Studying Animal Behavior: Autobiographies of the Founders.* Chicago: University of Chicago Press, 1985.

Doordan, Dennis P., ed. *Design History: An Anthology.* Cambridge: MIT Press, 1996.

Dormer, Peter. *The Art of the Maker: Skill and its Meaning in Art, Craft and Design.* New York: Thames & Hudson, 1994.

————, ed. *The Culture of Craft.* New York: Manchester University Press, 1997.

Dorner, Dietrich. *The Logic of Failure: Recognizing and Avoiding Error in Complex Situations.* Reading, Mass.: Addison-Wesley, 1996.

Dost, Martin H. "Simulation-A Part of Life at IBM." *Simulation,* November 1977, 188.

Dowdell, Chuck, and George Nicholas. "Tomorrow's Simulation Requirements." *Journal of Electronic Defense* 17, no. 7 (1994): 40–45.

Droz, Dan. "Prototyping: A Key to Managing Product Development." *Journal of Business Strategy* 13, no. 3 (1992): 34–38.

Duke, Richard D. "A Paradigm for Game Design." *Simulation & Gaming* 11, no. 3 (1980): 364–377.

Dunbar, Robin. *Grooming, Gossip and the Evolution of Language.* Cambridge: Harvard University Press, 1996.

————. *The Trouble with Science.* Cambridge: Harvard University Press, 1995.

Ebert, Roger. "Fatal Decision." *NY Post,* 3 October 1987.

Ehn, Pelle. *Work-Oriented Design of Computer Artifacts.* Stockholm: Arbetslivscentrum, 1988.

Epstein, Joshua M., and Robert Axtell. *Growing Artificial Societies: Social Science from the Bottom Up.* Cambridge: Brookings Institution Press/MIT Press, 1996.

Ferguson, Eugene S. *Engineering and the Mind's Eye.* Cambridge: MIT Press, 1992.

Fleming, Charles, "Testing. Testing, Why the Testing?" *Variety,* 17 February 1992.

Fleming, David. "Design Talk: Constructing the Object in Studio Conversation." *Design Issues* 14, no. 2 (1998).

Flores, Fernando, and Terry Winograd. *Understanding Computers and Cognition.* Norwood, N.J.: Ablex, 1986.

Floyd, C., et al., eds. *Software Development and Reality Construction*. New York: Springer-Verlag, 1992.

French, Michael. *Invention and Evolution: Design in Nature and Engineering*, 2d ed. New York: Cambridge University Press, 1994.

Friedhoff, Richard Mark, and William Benzon. *Visualization*. New York: Harry Abrams, 1989.

Fripp, John, "Why Use Business Simulations? *Executive Development* 7, no. 1 (1994): 29–32.

Fyfe, Gordon, and John Law, eds. *Picturing Power: Visual Depictions and Social Relations*. New York: Routledge, 1988.

Gagliardi, Pasquale, ed. *Symbols and Artifacts: Views of the Corporate Landscape*. Berlin: Walter de Gruyter, 1990.

Galbraith, Craig S., and Gregory B. Merrill. "The Politics of Forecasting: Managing the Truth." *California Management Review* 38, no. 2 (Winter 1996).

Galison, Peter. *Image and Logic: A Material Culture of Microphysics*. Chicago: University of Chicago Press, 1997.

Gass, Saul I. "Model World: A Model Is a Model Is a Model Is a Model." *Interfaces* 19, no. 3 (1989): 58–60.

Gauthier, Wendell. "Mock Juries, Shadow Jury, Community Survey Help." *The National Law Journal*, 13 February 1989, S16.

Gershefski, George W. "Building a Corporate Financial Model." *Harvard Business Review*, July–August 1969, 61–72.

———. "Corporate Models: The State of the Art." *Harvard Business Review*, Nov.–Dec. 1969, 1–6.

Ghislein, Brewster, ed. *The Creative Process: Reflections on Invention in the Arts and Sciences*. Berkeley: University of California Press, 1980.

Gilbert, Nigel, and Rosana Conte, eds. *Artificial Societies: The Computer Simulation of Social Life*. London: UCL Press, 1995.

Gilder, George. "Into the Fibersphere." *Forbes ASAP*, 7 December 1992.

Glasgow, Janice, N. Hari Narayanan, and B. Chandrasekaran, eds. *Diagrammatic Reasoning: Cognitive and Computational Perspectives*. Cambridge: MIT Press, 1995.

Goel, Vinod. *Sketches of Thought*. Cambridge: MIT Press, 1995.

Goffman, Erving. *Behavior in Public Places: Notes on the Social Organization of Gatherings*. New York: Free Press, 1963.

———. *Interaction Ritual*. New York: Doubleday, 1967.

———. *The Presentation of Self in Everyday Life*. New York: Doubleday, 1959.

Golovin, Lewis B. "Product Blending: A Simulation Case Study in Double Time: An Update." *Interfaces* 15, no. 4 (1985): 39–40.

Goodman, Jane, Edith Greene, and Elizabeth F. Loftus. "Runaway Verdicts or Reasoned Determinations." *Jurimetrics Journal of Lay Science and Technology*, Spring 1989, 285–309.

Gorb, Peter, ed. *Design Talks! London Business School Design Seminars.* London: The Design Council, 1988.

Green, Thad B., Sang M. Lee, and Walter B. Newsom. *The Decision Science Process: Integrating the Quantitive and Behavioral.* Petrocelli Books, 1978.

Greenbaum, Joan, and Morton Kyng. *Design at Work: Cooperative Design of Computer Systems.* Hillsdale, N.J.: Lawrence Erlbaum Associates, 1991.

Greenberger, Martin. "Simulation and the Problem of Air Traffic Control." *Machine Design* (1978), 27–43.

———. "A Way of Thinking about Model Analysis." *Interfaces* 10, no. 2 (1980).

Greenblat, Cathy Stein. "Group Dynamics and Game Design." *Simulation & Gaming* 11, no. 1 (1980): 35–58.

Grinyer, Peter H., and Jeff Wooller. "An Overview of a Decade of Corporate Modelling in the UK." *Accounting and Business Research* 11 (Winter 1980): 41–49.

Hall, William K. "Strategic Planning Models: Are Top Managers Really Finding Them Useful?" *Journal of Business Policy* 3, no. 2 (1973): 33–42.

Hannon, William J. "The Artist's Rendering." *Design Management Journal*, Spring 1995, 46–51.

Hardgrave, Bill C., and Rick L. Wilson. "An Investigation of Guidelines for Selecting a Prototyping Strategy." *Journal of Systems Management*, April 1994, 28–35.

Harre, Rom. *Great Scientific Experiments.* New York: Oxford University Press, 1983.

Hayes, R. H., and R. L. Nolan. "What Kind of Corporate Modelling Functions Best?" *Harvard Business Review*, May–June 1974, 51–68.

Hedstrom, Peter, and Richard Swedborg, eds. *Social Mechanisms: An Analytical Approach to Social Theory.* New York: Cambridge University Press, 1998.

Henderson, Kathryn. *On Line and On Paper: Visual Representations, Visual Culture and Computer Graphics in Design Engineering.* Cambridge: MIT Press, 1999.

Hequet, Marc. "Games that Teach." *Training* 32, no. 7 (1995): 53.

Herskovitz, Don. "All the World's a Stage: Future Trends in Simulations Systems." *Journal of Electronic Defense* 18, no. 5 (1995): 44(5).

Heudin, Jean-Claude, ed. *Virtual Worlds: First International Conference Proceedings.* New York: Springer-Verlag, 1998.

Higgins, James M. "Storyboard Your Way to Success." *Training and Development,* June 1995, 13–17.

Hiltzik, Michael. *Dealers of Lightning: Xerox PARC and the Dawn of the Computer Age.* New York: HarperCollins, 1999.

Hindle, Brook. *Emulation and Invention.* New York: NYU Press, 1981.

Hirshberg, Jerry. *The Creative Priority: Driving Innovative Business in the Real World.* New York: HarperBusiness, 1998.

———. "Designing the Soul of the Machine." *I.D.* Jan./Feb. 1993, 54.

Holland, John H. *Emergence: From Chaos to Order.* Reading, Mass.: Addison-Wesley, 1998.

Holst, Per A. "Simulation at Foxboro—A Look Back Over 40 Years." *Simulation,* November 1977, 173–176.

Horgan, John. *The End of Science: Facing the Limits of Knowledge in the Twilight of the Scientific Age.* Reading, Mass.: Addison-Wesley, 1996.

"How to Make Concurrent Engineering Work." Parts 11, 13, 15, 16. *Machine Design.* Special Series, 1995.

Howard, Bennet. "Manifesto for a Theory of Drama and Irrational Choice." *Systems Practice* 6: 492–494.

Howard, Kenneth R. "Unjamming Traffic with Computers." *Scientific American,* October 1997.

Howie, John. "Bringing Case Issues Into Focus: Mock Trials and Focus Groups Can Sharpen Your Presentation." *Trial,* January 1995, 32.

Huizinga, John. *Homo Ludens: A Study of the Play Element in Culture.* Boston: Beacon Press, 1955.

Hutchinson, David. "Storyboarding Special Effects." *Starlog,* April 1981, 20–23.

Hyams, Joe. "The Rehearsed Picture—Will It Be Better?" *NY Herald Tribune,* 11 June 1961.

Iansiti, Marco. *Technology Integration: Making Critical Choices in a Dynamic World.* Boston: Harvard Business School Press, 1998.

Imagineers. *Imagineering.* New York: Hyperion, 1996.

Jeffries, Jim. "Storyboard to the Marketing Budget." *Executive,* July/August 1996, 18–24.

Jewell, P., and Caroline Loizos. *Play, Exploration, and Territory in Mammals.* New York: Academic Press, 1966.

John-Steiner, Vera. *Notebooks of the Mind.* New York: Harper & Row, 1985.

Jones, Chuck. *Chuck Reducks: Drawing from the Fun Side of Life.* New York: Warner Books, 1996.

Judson, Horace Freeland. *The Eighth Day of Creation: The Makers of the Revolution in Biology.* New York: Simon & Schuster, 1979.

Kanigel, Robert. *Apprentice to Genius.* New York: Macmillan, 1986.

Keen, Peter G. W. *The Process Edge: Creating Value Where It Counts.* Boston: Harvard Business School Press, 1997.

Keeprow, Alice. "Can You Sketch on a Computer?" *International Design,* Jan./Feb. 1993, 81–83.

Kelly, Kevin. *Out of Control.* Reading, Mass.: Addison-Wesley, 1995.

Kennedy, Biff. "The Models That Didn't Work." *Interfaces* 8, no. 2 (1978): 54.

Keough, Mark, and Andrew Doman. "The CEO as Organization Designer." *The McKinsey Quarterly* no. 2 (1992). Interview with Jay Forrester.

Kidder, Tracy. *The Soul of a New Machine.* New York: Atlantic Monthly Press, 1981.

Kitchener, Alan. "The Impact of Technology on the Information Systems and OR Profession." *Interfaces* 16, no. 3 (1986): 20–30.

Kitfield, James. "Simulation Technologies Merge in Virtual Battlefield." *Government Executive* 27, no. 8 (1995): 142–144.

Koch, Neal. "She Lives! She Dies! Let the Audience Decide." *New York Times,* 19 April 1992.

Koch, Richard. *The 80/20 Principle.* New York: Currency Doubleday, 1998.

Kody, Kori. "Storyboarding as a Basis for Collaboration." *Technical Communication,* first quarter 1992, 83.

Koselka, Rita. "Businessman's Dilemma." *Forbes,* 11 October 1993, 107.

——— . "Games Businesses Play," *Forbes,* 7 November 1994, 12.

Krause, Martin, and Linda Witkowski. *Walt Disney's Snow White and the Seven Dwarfs: An Art in Its Making.* New York: Hyperion, 1994.

Krugman, Paul. *Development, Geography and Economic Theory.* Cambridge: MIT Press, 1997.

Kyng, Morten, and Lars Mathiassen, eds. *Computers and Design in Context.* Cambridge: MIT Press, 1997.

Lane, David A.,"Models and Aphorisms." Sante Fe Institute, Santa Fe, N.M., 1994.

Lane, David C. "Modelling as Learning." *European Journal of Operational Research* 59 (1992): 64.

——— . "The Road Not Taken: Observing a Process of Issue Selection and Model Concept." *Systems Dynamics Review* 9:239–264.

Lane, David, and Robert Maxfield. "Foresight, Complexity, and Strategy." Santa Fe Institute, Santa Fe, N.M., 16 December 1995.

Langham, Barbara A. "Drawing It Out." *Successful Meetings,* January 1994, 114–117.

Lave, Jean. *Cognition in Practice: Mind, Mathematics and Culture in Everyday Life.* New York: Cambridge University Press, 1988.

Lawson, Bryan. *Design in Mind.* London: Butterworth Architecture, 1994.

———. *How Designers Think: The Design Process Demystified.* London: Butterworth Architecture, 1980.

Learch, Francisco Javier. "Computerized Financial Planning: Discovering Cognitive Difficulties in Model Building." University of Michigan, 1988, 281.

Lederman, Linda Costigan. "Debriefing: Toward a Systematic Assessment of Theory and Practice." *Simulation & Gaming* 23, no. 2 (1992): 145–160.

———. "Guest Editorial: After the Game Is Over." *Simulation & Gaming* 23, no. 2 (1992): 143–144.

Lee, Youngho, and Amie Elcan. "Simulation Modeling for Process Reengineering in the Telecommunications Industry." *Interfaces* 26, no. 3 (1996): 1–9.

Leon, Linda, et al. "Spreadsheets and OR/MS Models: An End-User Perspective." *Interfaces* 26, no. 2 (1996): 92–104.

Leonard, J. Barry. "Assessing Risk Systematically." *Risk Management,* January 1995, 12–17.

Leonard-Barton, Dorothy. *Wellsprings of Knowledge: Building and Sustaining the Sources of Innovation.* Boston: Harvard Business School Press, 1995.

Liker, Jeffrey, John E. Ettlie, and Jon C. Campbell, eds. *Engineered in Japan: Japanese Technology-Management Practices.* New York: Oxford University Press, 1995.

Little, J. D. C. "Models and Managers: Concept of a Decision Calculus." *Management Science,* April 1970, B69–87.

Lloyd, Tom. "Management: The Games People Play." *Financial Times,* 1 March 1995, 17.

Lu, Jingxi, and Biao Lu. "Non-Hedgeable Risk: Model Risk." *Capital Market Risk Advisors' White Paper,* New York, 1997.

Lubar, Steven, and W. David Kingery, eds. *History from Things: Essays on Material Culture.* Washington, D.C.: Smithsonian Institution Press, 1993.

Lupton, Ellen. "Interpreting Design's Visual Language." *I.D.,* January/February 1993, 60.

Lutz, Robert A. *Guts: The Seven Laws of Business That Made Chrysler the World's Hottest Car Company.* New York: John Wiley & Sons, 1998.

MacCready, Paul, and John S. Langford. "Human Powered Flight: Perspectives on Processes and Potentials." The 28th Lester D. Gardner Lecture. Presented at MIT, Cambridge, Mass., 1998.

Machol, Robert E. "Thirty Years of Modelling Midair Collisions." *Interfaces* 25, no. 5 (1995): 151–172.

MacKenzie, Donald. *Knowing Machines*. Cambridge: MIT Press, 1996.

Malone, Michael S. *The Big Score: The Billion Dollar Story of Silicon Valley*. New York: Doubleday, 1985.

Maltin, Leonard. *Of Mice and Magic: A History of American Animated Cartoons*. New York: Plume, 1987.

Manning, Phillip. *Erving Goffman and Modern Sociology*. Stanford, Calif.: Stanford University Press, 1992.

Margolin, Victor, and Richard Buchanan, eds. *The Idea of Design: A 'Design Issues' Reader*. Cambridge: MIT Press, 1996.

Marling, Karel Ann. *Designing Disney's Theme Parks: The Architecture of Reassurance*. Flammarion, 1997.

Marriott, Willard, and Bob Cross. "Room at the Inn: Revenue Management." *CEO Magazine*, July 1997, 23–27.

MacCormack, Alan, and Marco Iansiti. "Team New Zealand (A)" and "Team New Zealand (B)." Harvard Business School Case Study 9-697-040 and 9-697-041.

Marshall, Chris, Larry Prusak, and David Shpilberg. "Financial Risk and the Need for Superior Knowledge Management," *California Management Review* 38, no. 3 (Spring 1996): 77–101.

Masi, C. G. "Putting 'Quiet Shoes' on Household Appliances." *R&D* 38, no. 12 (1996): 20–26.

Mason, Richard O., and Ian I. Mitroff. "Assumptions of Majestic Metals: Strategy Through Dialectics." *California Management Review* 22, no. 2 (1979): 80–88.

McCloud, Scott. *Understanding Comics*. Toronto: Tundra Publishing, 1993.

McConnell, Steve. *Rapid Development: Taming Wild Software Schedules*. Redmond, Wash.: Microsoft Press, 1996.

McCullough, Malcolm. *Abstracting Craft: The Practiced Digital Hand*. Cambridge: MIT Press, 1996.

McHaney, Roger Wilson. "Recurrent Factors in Computer Simulation Success: An Empirically Tested Contingency Model." Ph.D. diss., University of Arkansas, 1995.

McLean, Ephraim, and Gary Neale. "Computer-Based Planning Models Come of Age." *Harvard Business Review*, July–August 1980, 46–48.

McNeil, Margaret. "Industry Simulation: A New Type of Business Game Tapping Both Analytic and Synthetic Skills." *Training and Development Methods* 9, no. 3 (1995): 6.27–6.39.

Miller, Jonathan. *Subsequent Performances*. New York: Elisabeth Sifton Books/Viking, 1986.

Mitchell, C. Thomas. *Redefining Designing: From Form to Experience*. New York: Van Nostrand Reinhold, 1993.

Mithen, Steven. *The Prehistory of the Mind: The Cognitive Origins of Art, Religion and Science*. New York: Thames & Hudson, 1996.

Mokyr, Joel. *The Lever of Riches: Technological Creativity and Economic Progress*. New York: Oxford University Press, 1990.

Morecroft, John. "Introduction and Background." *European Journal of Operational Research* 59 (1992): 6–8.

Morecroft, John D. W., and John. D. Sterman, eds. *Modeling for Learning Organizations*. Portland, Ore.: Productivity Press, 1994.

Morecroft, John D. W. and Kees A. J .M. van der Heijden. "Modelling the Oil-Producers-Capturing Oil Industry Knowledge in a Behavioral Simulation Model." *European Journal of Operational Research* 59 (1992): 102–122.

Morris, W. T. "On the Art of Modelling." *Management Science*, August 1967, B-707–B-717.

Murray, Charles J. "Chrysler's Digital Trailblazer: The Redesigned LH Vehicles Break New Ground in CAD and Assembly." *Design News*, 6 October 1997.

Nachmanovitch, Stephen. *Free Play: Improvisation in Life and Art*. New York: Jeremy P. Tarcher, 1990.

Nansus, J. "Management Games: An Answer to Critics." In *Business Games Handbook*. R. G. Graham and C. E. Gray, eds. 52–57. New York: American Management Association, 1969.

Naylor, T. H. "Effective Use of Strategic Planning, Forecasting, and Modelling in the Executive Suite." *Managerial Planning*, January/February 1982, 4–11.

———. "Strategic Planning Models." *Managerial Planning*, July/August, 3–11.

Naylor, Thomas H. "Why Corporate Planning Models?" *Interfaces*. (November 1977): 87–94.

Naylor, Thomas H., and Daniel R. Gattis. "Corporate Planning Models." *California Management Review* 18, no. 4 (Summer 1976): 69.

Naylor, T., and J. Schailand. "A Survey of Users of Corporate Planning Models." *Management Science*, May 1976, 930.

Nishiguchi, Toshiro, ed. *Managing Product Development*. New York: Oxford University Press, 1996.

Nonaka, Ikujiro, and Hiro Takeuchi. *The Knowledge-Creating Company.* New York: Oxford University Press, 1995.

Nordwall, Bruce D. "Aerospace Advances Challenge Training." *Aviation Week & Space Technology,* 2 September 1996.

———. "Military Orders Outpace Civil Demand for Simulators." *Aviation Week & Space Technology,* 2 September 1996.

Norman, Donald A. *The Design of Everyday Things.* New York: Doubleday Currency, 1988.

———. *The Invisible Computer.* Cambridge: MIT Press, 1998.

———. *Things That Make Us Smart.* Reading, Mass.: Addison-Wesley, 1993.

Norton, Rob. "Winning the Game of Business." *Fortune,* 6 February 1995.

Oreskes, Naomi, Kenneth Belitz, and Kristin Shrader. "Verification, Validation and Confirmation of Numerical Models in the Earth Sciences." *Science,* 4 February 1994, 641–646.

Pagels, Heinz R. *The Dreams of Reason: The Computer and the Rise of the Sciences of Complexity.* New York: Bantam Books, 1989.

Parkinson, Chris. "What If? Decision Shaping Systems." *CMA—The Management of Accounting Magazine,* 69, no. 2 (1995): 10.

Patton, Phil. "Designing for the Soul of the Machine." *International Design,* Jan./Feb. 1993, 54.

———. "Dipping into the Automotive Gene Pool." *International Design,* Jan./Feb. 1993, 60.

Pellegrini, Anthony D., ed. *The Future of Play Theory.* SUNY Press, 1995.

Perla, Peter P. *The Art of Wargaming.* Annapolis, Md.: Naval Institute Press, 1990.

Petroski, Henry. *Engineers of Dreams: Great Bridge Builders and the Spanning of America.* New York: Vintage Books, 1996.

———. *The Evolution of Useful Things.* New York: Vintage Books, 1992.

———. *Invention by Design: How Engineers Get from Thought to Thing.* Cambridge: Harvard University Press, 1996.

Phelan, Richard J., and Mary Patricia Benz. "Mock Trials: Experiments That Work." *The National Law Journal,* 29 July 1985, 15ff.

Pickthall, Barry. "IACC Worlds Focus Attention in Design." *Yachting* 177, no. 1 (1995): 16.

Pipes, Alan. *Drawing for 3-Dimensional Design: Concepts Illustration Presentation.* Dubuque, Iowa: William C. Brown Publishers, 1992.

Poole, Trevor. *Social Behavior in Mammals.* New York: Chapman & Hall, 1985.

Porter, Tom. *How Architects Visualize.* New York: Van Nostrand Reinhold, 1979.

Poundstone, William. *Prisoner's Dilemma: John von Neumann, Game Theory and the Puzzle of the Bomb.* New York: Doubleday, 1992.

Powell, Dick. "Take Two: A Dissection of the Design Process." *Design,* April 1993, 20–25.

Pressman, Andy. *The Fountainheadache: The Politics of Architect-Client Relations.* New York: John Wiley & Sons, 1995.

Prietula, Michael J., Kathleen M. Carley, and Les Gasser, eds. *Simulating Organizations: Computational Models of Institutions and Groups.* Cambridge: AAAI Press/MIT Press, 1998.

Pye, David. *The Nature of Design.* New York: Van Nostrand Reinhold, 1964.

Raser, John R. *Simulation and Society: An Exploration in Scientific Gaming.* Boston: Allyn & Bacon, 1969.

Rechtin, Eberhardt. *System Architecting: Creating & Building Complex Systems.* New York: Prentice-Hall, 1991.

Rechtin, Eberhardt, and Mark W. Maier. *The Art of Systems Architecting.* Boca Raton, Fla.: CRC Press, 1997.

Reinertsen, Donald G. *Managing the Design Factory.* New York: Free Press, 1997.

Reiss, David. "Making the Impossible Come True." *Filmmakers Monthly,* June 1980, 21–34.

Rheingold, Howard. *Tools for Thought: The History and Future of Mind-Expanding Technology.* New York: Simon & Schuster, 1985.

Rheingold, Howard, and Howard Levine. *The Cognitive Connection: Thought and Language in Mind and Machine.* New York: Prentice-Hall, 1987.

Rhodes, Richard. *The Making of the Atomic Bomb.* New York: Simon & Schuster, 1986.

Richels, Richard. "Building Good Models Is Not Enough." *Interfaces* 11, no. 4 (1981): 48–54.

Riordan, Michael, and Lillian Hoddeson. *Crystal Fire: The Birth of the Information Age.* New York: W.W. Norton, 1997.

Roberts, Al. "What's Wrong with Business Games?" In *Business Games Handbook.* R. G. Graham and C. E. Gray, eds. 47–51. New York: American Management Association, 1969.

Robson, Andrew J. "The Spreadsheet: How It Has Developed Into a Sophisticated Modelling Tool." *Logistics Information Management* 7, no. 1 (1994): 17–23.

Robson, David. "More Vision, Less Dazzle." *Design,* May 1993, 30–33.

Romberg, Sigmund. "Plays Are Made In Rehearsal." *New York Times,* 10 February 1927.

Root-Bernstein, Robert Scott. *Discovering*. Cambridge: Harvard University Press, 1989.

Rosen, Stephen Peter. *Winning the Next War: Innovation and the Modern Military*. Ithaca, NY: Cornell University Press, 1991.

Rosenbloom, Richard S., and William J. Spencer, eds. *Engines of Innovation: U.S. Industrial Research at the End of an Era*. Boston: Harvard Business School Press, 1996.

Rosenthal, Larry. "Can You Sketch on a Computer?" *I.D.*, January/February 1993, 81–83.

Rudd, Jim, Ken Stern, and Scott Isensee. "Low vs. High-Fidelity Prototyping Debate." *Interactions* (January 1996): 77–85.

Russell, Joyce E. A. "Flying Fox: A Business Adventure in Teams and Team-work," *Personnel Psychology* 47, no. 3 (1994): 688–691.

Saaty, Thomas L., and F. Joachim Weyl, eds. *The Spirit and Uses of the Mathematical Sciences*. New York: McGraw-Hill, 1969.

Sabbagh, Karl. *Twenty-First Century Jet: The Making and Marketing of the Boeing 777*. New York: Scribner, 1996.

Salamone, Salvatore. "What's the Story?" *Byte*, March 1996, 67–70.

Salton, Gary. "Human Engineering." *Managerial Planning*, March/April, 1982, 31–35.

Sanoff, Henry. *Visual Research Methods in Design*. New York: Van Nostrand Reinhold, 1991.

Saunder, Danny, Fred Percival, and Mati Vartianen, eds. *Games and Simulations to Enhance Quality Learning*. Vol. 4 of *The Simulation and Gaming Yearbook*. London: Kogan Page, 1996.

Scerbo, Mark W., and Mustapha Mouloua, eds. *Automation and Human Performance: Current Research and Trends*. Hillside, N.J.: Lawrence Erlbaum Associates, 1999.

Schefter, James. *All Corvettes are Red: The Rebirth of an American Legend*. New York: Simon & Schuster, 1996.

Schelling, Thomas C. *Choice and Consequence: Perspectives of an Errant Economist*. Cambridge: Harvard University Press, 1984.

Schlesinger, Stewart I. "The Simulation Experience As Viewed at the Aerospace Corporation." *Simulation*, November 1977, 179–180.

Schön, Donald A. *Educating the Reflective Practitioner*. San Francisco: Jossey-Bass, 1991.

———. *The Reflective Practitioner: How Professionals Think in Action*. New York: Basic Books, 1983.

Schrage, Michael. *No More Teams!: Mastering the Dynamics of Creative Collaboration*. New York: Doubleday Currency, 1995.

————. *Shared Minds: The New Technologies of Collaboration.* New York: Random House, 1990.

Schumacher, Joseph E., and Stanley L. Brodsky. "The Mock Trial: An Exploration in Interdisciplinary Training." *Law and Psychology Review,* Spring 1988, 79–93.

"Seeing Themselves As Others See Them," *London Times,* 5 November 1932.

Shirreff, David. "Lessons from NatWest." *Euromoney,* May 1997, 42–47.

————. "The Rise and Rise of the Risk Manager." *Euromoney,* February 1998, 56–60.

Showers, Jack Paul. "Easy Mock Trials . . . So You Don't Have to Ask Why?" *Louisiana Bar Journal,* April 1993, 605.

Simon, Herbert A. *Administrative Behavior.* New York: Macmillan, 1945.

————. *The Sciences of the Artificial,* 2d ed. Cambridge: MIT Press, 1981.

Simulation & Gaming: An International Journal of Theory, Practice and Research. Silver Anniversary Edition, pts. 1, 2, and 3 (1994–1995). Newbury Park, Calif.: Sage Periodicals Press.

Sisodia, Rajendra S. "Competitive Advantage Through Design." *Journal of Business Strategy* 13, no. 6 (1992): 33–40.

Slack, Kim. "Training for the Real Thing." *Training & Development* 47, no. 5 (1993): 79.

Sloan, Alfred P., Jr. *My Years with General Motors.* New York: Doubleday, 1990.

Solomon, Charles. *The Disney That Never Was.* New York: Hyperion, 1995.

Steiner, Gary A., ed. *The Creative Organization.* Chicago: University of Chicago Press, 1965.

Steinherr, Alfred. *Derivatives: The Wild Beast of Finance.* New York: John Wiley & Sons, 1998.

Stodgill, Ralph M., ed. *The Process of Model-Building in the Behavioral Sciences.* Columbus, Ohio: Ohio State University Press, 1970.

Sudjic, Deyan. *Cult Objects.* London: Paladin Books, 1985.

Sullivan, R. Lee. "Virtual Reality Made Easy." *Forbes* 155, no. 1 (1995): 66(2).

Suri, Rajan, et al. "From CAN-Q to MPX: Evolution of Queuing Software for Manufacturing," *Interfaces* 25, no. 5 (1995): 128–149.

Sutton, Robert I., and Thomas A. Kelley. "Creativity Doesn't Require Isolation: Why Product Designers Bring Visitors 'Backstage.'" *California Management Review* 40, no. 1 (Fall 1997): 75–91.

Sutton-Smith, Brian. *The Ambiguity of Play.* Cambridge: Harvard University Press, 1997.

Swart, William, and Luca Donno. "Simulation Modelling Improves Operations, Planning, and Productivity of Fast Food Restaurants." *Interfaces* 11, no. 6 (1981).

Swift, E. M. "Bok to the Future." *Sports Illustrated*, 3 July 1995, 32.

Taha, Hamdy, and Pable Nuno de la Parra. "A SIMNET Simulation Model for Estimating System." *Computers & Industrial Engineering* 17, no. 1–4 (1989): 317–322.

Taylor, Keith. "Kiwis Take Tough Line on Fairness." *Yachting* 178, no. 2 (1995): 26–27.

———. *Yachting*. July 1995 (39–45, 56, 60); June 1995 (58, 74); January 1995 (16–17).

Thackara, John. *Design after Modernism*. New York: Thames & Hudson, 1988.

Thiagarajan, Sivasailam. "How I Designed a Game—And Discovered the Meaning of Life." *Simulation & Gaming*, December 1994, 529–539.

Thomas, Frank, and Ollie Johnston. *The Illusion of Life: Disney Animation*. New York: Hyperion, 1981.

Thomke, Stefan H. "Managing Experimentation in the Design of New Products." *Management Science* 44, no. 6 (1998).

Thomke, Stefan. "Simulation, Learning and R&D Performance: Evidence from Automotive Development." *Research Policy* 27 (1998): 55–74.

Thomke, Stefan, and Takahiro Fujimoto. "The Effect of 'Frontloading' Problem Solving on Product Development Performance." Working paper, Harvard Business School, Boston, December 1998.

Tokaty, G. A. *A History and Philosophy of Fluid Mechanics*. Mineola, N.Y.: Dover Publications, 1971.

Tompkins, Joanne. "The Story of Rehearsal Never Ends." *Canadian Literature* 144 (spring 1995): 142–161.

Ulrich, Karl T., and Steven D. Eppinger. *Product Design and Development*. New York: McGraw-Hill, 1995.

Valentine, Don. Interview by Andree Abecassis. In *Inc.*, May 1985, 29.

Vennix, Jac A. M. *Group Model Building: Facilitating Team Learning Using System Dynamics*. New York: John Wiley & Sons, 1996.

Vertrees, Alan David. "Reconstructing the 'Script in Sketch Form': An Analysis of the Narrative Construction and Production Design of the Fire Sequence in Gone With the Wind." *Film History* 3 (1989): 87–104.

Vincenti, Walter G. *What Engineers Know and How They Know It: Analytical Studies from Aeronautical History*. Baltimore: Johns Hopkins University Press, 1993.

Vinson, Donald E. "The Shadow Jury: An Experiment in Litigation." *ABA Journal*, October 1982, 1242–1246.

Wallace, William A., ed. *Ethics in Modeling*. New York: Pergamon Press, 1994.

Walsh, Mike. "Two Steps Forward One Step Back." *Infosystems*, January 1988, 52–53.

Walton, Mary. *Car: A Drama of the American Workplace*. New York: W.W. Norton, 1997.

Watson, James D. *The Double Helix*. New York: W.W. Norton, 1980.

Wegner, Peter. "Why Interaction Is More Powerful than Algorithms." *Communications of the ACM* 40, no. 5: 80.

Weick, Karl E. *The Social Psychology of Organizing*, 2d ed. New York: Random House, 1979.

Weintraub, E. Roy, ed. *Toward a History of Game Theory*. Durham, N.C.: Duke University Press, 1992.

Weisberg, Robert W. *Creativity: Genius and Other Myths*. New York: W. H. Freeman, 1986.

Wenger, Etienne. *Communities of Practice: Learning, Meaning and Identity*. New York: Cambridge University Press, 1998.

Whitehead, Alfred North. *Science and the Modern World*. New York: Macmillan, 1925.

Wiender, Earl, Barbara G. Kanki, and Robert L. Helmreich, eds. *Cockpit Resource Management*. New York: Academic Press, 1993.

Williams, Gaynor. "Convenient Solutions." *Design*, May 1993, 14–18.

Wilmer, Harry A., ed. *Creativity: Paradoxes and Reflections*. Wilmette, Ill.: Chiron Publications, 1991.

Winner, Langdon. "Do Artifacts Have Politics?" *Daedalus* 109 (1980): 121–136.

Winograd, Terry, ed. *Bringing Design to Software*. Reading, Mass.: ACM Press/Addison-Wesley, 1996.

Wolfe, Joseph. "A Forecast of Business Gaming in the Year 2010." *Simulation & Gaming* 24, no. 3 (1993): 373–375.

Wolpert, Lewis. *The Unnatural Nature of Science*. Cambridge: Harvard University Press, 1992.

Wolpert, Lewis, and Alison Richards. *Passionate Minds: The Inner World of Scientists*. New York: Oxford University Press, 1997.

Wood, Gerry. "Using Simulation for Process Improvement." *Management Services* 37, no. 10 (1993): 18–19.

Woolsey, Kristina Hooper, Scott Kim, and Gayle Curtis. *VizAbility*. Boston: PWS Publishing, 1996.

Wright, Orville. *How We Invented the Airplane: An Illustrated History*. Mineola, N.Y.: Dover, 1988.

Yates, Brock. *The Critical Path: Inventing an Automobile and Reinventing a Corporation*. Boston: Little, Brown, 1996.

Yelavich, Susan, ed. *The Edge of the Millennium: An International Critique of Architecture, Urban Planning Product and Communication Design.* Whitney Library of Design, 1993.

Zeckhauser, Richard J., Ralph L. Keeney, and James K. Sebenius. *Wise Choices: Decisions, Games, and Negotiations.* Boston: Harvard Business School Press, 1996.

Zeruvabel, Eviator. *Social Mindscapes: An Invitation to Cognitive Sociology.* Cambridge: Harvard University Press, 1997.

I N D E X

A B O U T   T H E   A U T H O R

Michael Schrage is a Research Associate with the MIT Media Lab and a Merrill Lynch Forum Innovation Fellow. His ongoing work explores the role of prototypes and simulations in risk management and innovation. His clients have included McKinsey & Co., IBM, Procter & Gamble, NTT, Hewlett-Packard, Andersen Consulting, International Thomson, Calico Commerce, and General Motors. He writes the fortnightly "Brave New Work" column for *Fortune* magazine. His last book, *No More Teams: Managing the Dynamics of Creative Collaboration*, was published in 1995.